REVIEW FOR THE MATHEMATICS SECTION OF THE GED TEST

By
ROBERT FLOYD

This book is correlated to the videotape produced by COMEX Systems, Inc., Review for The GED Mathematics test by Robert Floyd ©2001 it may be obtained from

comex systems, inc.
5 Cold Hill Rd.
Suite 24
Mendham, NJ 07945

Published by

comex systems, inc.

5 Cold Hill Rd. South, Suite 24
Mendham, NJ 07945

ISBN 1-56030-143-0

Table of Contents

Introduction

The current series of the GED test includes major changes from the older series of the test. The mathematics section of the test contains some of the biggest changes.

The Mathematics test is now divided into two parts. In the first part you may use a calculator. The calculator will be provided to you on test day and everyone will be using the same type of calculator.

The second major change to the test concerns the format of the questions. The questions are no longer all in the multiple-choice format. The test contains approximately 20% alternate format questions. The first type of alternate format question is a grid. In this type of question if the answer is 99 you will have to grid in a circle containing 9 in two columns. The second type of alternate response is a coordinate plane where you are expected to darken in the circle that represents a point you have calculated. Both of these alternate format questions are discussed in Appendix B.

The test questions cover the topics: Number Properties; Algebra and Graphing; Geometry and Measurement; Probability and Statistics. The approximate percent of each question is found in the table below. About 50% of the questions will be related to a graphic, and 25% of the questions will be found in problem sets.

	PROCEDURAL	CONCEPTUAL	APPLICATIONS MODELING PROBLEM SOLVING
Number Properties	5%	7.5%	12.5%
Measurement & Geometry	5%	7.5%	12.5%
Data Analysis Statistics	5%	7.5%	12.5%
Algebra	5%	7.5%	12.5%

The questions will all be word problems and they will deal with everyday life topics. For example, you might have questions on a graph of home energy consumption or a percent problem figuring out the tip on a meal in a restaurant.

This book covers all of the mathematical concepts you will see on the exam. If you can work comfortably with fractions, decimals, and percents, you are well on your way to passing this section of the test.

Detailed Breakdown of Content Areas

Number Properties

- Work with whole numbers, integers, fractions, decimals, percents and exponents.
- Convert whole numbers, integers, fractions, decimals, percents and exponents.
- Analyze and apply ratios, proportions and percents.
- Correctly use order of operations.
- Apply the appropriate operations.
- Perform operations (with a calculator) using fractions, decimals and percents that have no restrictions on size or result.
- Use estimation to solve problems.

Measurement and Geometry

- Solve problems using perpendicularity, parallelism, congruence and similarity.
- Use Pythagorean Theorem and right angle trigonometry to solve problems.
- Find and use the slope and y intercept of a line.
- Convert units of measure.
- Solve problems involving perimeter, length, area, surface area, volume, angles, weight and mass.
- Use uniform rates.
- Interpret scales, meters and guages.
- Predict impact of changes in linear distance on changes in perimeter, area, and volume.

Data Analysis, Statistics and Probability

- Interpret tables, charts, and graphs.
- Make decisions based on data analysis.
- Evaluate arguments based on data analysis.
- Apply measures of central tendency (mean, mode, median).
- Make predictions from data.
- Recognize bias in statistical claims.
- Use and interpret a frequency distribution of outcomes.
- Compare sets of data based on central tendency and dispersion.

Algebra, Functions, and Patterns

- Analyze and represent situations involving variable quantities with tables, graphs, verbal descriptions and equations.
- Convert between different representations, such as tables, graphs, verbal descriptions and equations.
- Create and use algebraic equations.
- Evaluate formulas and functions.
- Solve equations and systems of equations.
- Use direct and indirect variation.
- Analyze tables and graphs to identify patterns and relationships.
- Explain how a change in one quantity will result in a change in another quantity.

Detailed Breakdown of the Cognitive Levels

Procedural

- Select and apply appropriate procedures.
- Justify the correctness of a procedure.
- Modify procedures for different problem settings.
- Use numerical algorithms.
- Read graphs and charts.
- Perform geometric constructions.
- Round numbers.
- Order lists of numbers.

Conceptual

- Recognize and label concepts.
- Generate examples.
- Interrelate models and diagrams.
- Identify and apply principles.
- Apply facts and definitions.
- Compare and contrast concepts and principles.
- Interpret and apply signs, symbols and terms.
- Interpret assumptions and relations.

Application, Modeling and Problem Solving

- Recognize and formulate problems.
- Determine the sufficiency and consistency of data.
- Use data, models and relevant math.
- Generate, extend, and modify procedures.
- Judge the reasonableness and correctness of solutions.

Key Point

A fraction expresses a relationship between two numbers – one written on top of the other, with a line in between them. The top number is called the **numerator**, which represents a part of the bottom number, called the **denominator**. The line that separates them is a symbol for division.

$$\frac{3}{8} = \frac{\text{numerator}}{\text{denominator}}$$

The figure below is divided into eight equal pieces. The fraction of the figure that is shaded is $\frac{3}{8}$. Eight represent the total number of pieces that the figure is divided into. Eight represents the whole figure. Three represents the part of the figure that is shaded. By placing the number that represents the part (3) over the number that represents the whole (8), we represent the part of the figure that is shaded as a fraction ($\frac{3}{8}$).

$$\text{Fraction Shaded} = \frac{\text{Pieces Shaded}}{\text{Total Pieces}} = \frac{3}{8}$$

As you work with fractions, you'll find that two fractions may very well be equal even though they are expressed in different forms. Therefore, it will be helpful for you to be able to change from one form into another.

Section 1: Reducing A Fraction To Lowest Terms

The figure below is divided into 12 equal parts. Four of the parts are shaded.

One way to express the fraction that is shaded is to places the number of shaded parts (4) over the total number of parts (12). This fraction is $\frac{4}{12}$.

If we consider the figure as a collection of six columns, 2 of which are shaded, we could express the fraction as the number of shaded columns (2) over the total number of columns (6). This fraction is $\frac{2}{6}$.

Now suppose we kept the same number of blocks and the same number of shaded blocks, but rearranged them to look like the figure below.

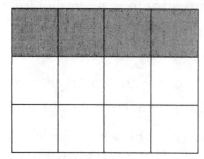

If we consider the figure as a collection of 3 rows, 1 of which are shaded, we could express the fraction as the number of shaded rows (1) over the total number of rows (3). This fraction is $\frac{1}{3}$.

All three fractions, $\frac{4}{12}$, $\frac{2}{6}$, and $\frac{1}{3}$ express the relationship between the 4 shaded boxes to the total number of boxes. These fractions have the same value, and we say that they are **equivalent**.

The relationship that is expressed in lowest terms is the fraction $\frac{1}{3}$. A fraction can be reduced if there is a whole number that can be divided evenly into both the numerator (top) and denominator (bottom). A fraction has been reduced to **lowest terms** if there is no longer any whole number, other than one, that can be divided evenly into both the numerator and denominator.

In our example, the fraction $\frac{4}{12}$ can be reduced to the fraction $\frac{2}{6}$ by dividing both the numerator and denominator by the whole number 2.

$$\frac{4 \div 2}{12 \div 2} = \frac{2}{6}$$

However, this fraction is not reduced to lowest terms because there is still a whole number that will divide evenly into 2 and 6. The number 2 will divide evenly into both the numerator and denominator.

$$\frac{2 \div 2}{6 \div 2} = \frac{1}{3}$$

We could have gotten the same result by dividing the numerator of the original fraction (4) and the denominator of the original fraction (12) each by the number 4.

$$\frac{4 \div 4}{12 \div 4} = \frac{1}{3}$$

Working with a fraction in its lowest terms will give you a better sense of its relative size. It will also help you add and subtract fractions, which we will cover in a later section.

Example: Reduce $\frac{36}{90}$ to lowest terms.

Solution: The largest number that can be divided evenly into 36 and 90 is 18, but chances are it will not be the first one you see. It is more likely that you would see that 9 divides evenly into both numbers. There is not just one correct way to start the problem. However, there is only one correct answer.

$$\frac{36 \div 9}{90 \div 9} = \frac{4}{10}$$

At this stage, you have reduced the fraction, but not to its lowest terms. Both 4 and 10 can be divided evenly by the number 2.

$$\frac{4 \div 2}{10 \div 2} = \frac{2}{5}$$

The answer is $\frac{2}{5}$. There is no whole number, other than one, that can be divided evenly into both 2 and 5.

Example: Reduce $\frac{20}{35}$ to lowest terms.

Solution: The largest number that can be divided evenly into 20 and 35 is 5.

$$\frac{20 \div 5}{35 \div 5} = \frac{4}{7}$$

Example: Reduce $\frac{175}{250}$ to lowest terms.

Solution: It might be helpful to think of these numbers in terms of dollars and cents. Ask yourself how many quarters are in $1.75? How

3

many quarters are there in $2.50? There are 7 quarters in $1.75 and 10 quarters in $2.50.

$$\frac{175 \div 25}{250 \div 25} = \frac{7}{10}$$

Example: Reduce $\frac{72}{108}$

Solution: $\frac{72 \div 36}{108 \div 36} = \frac{2}{3}$

The largest number that divides evenly into 72 and 108 is 36. Don't feel badly if you didn't initially identify 36. Maybe you saw 2, or maybe 9. Dividing both numerator and denominator by 2 would result in the fraction $\frac{36}{54}$. If you divided 72 and 108 by 9, you would end up with the fraction $\frac{8}{12}$. In both cases these fractions can still be reduced. As long as you continue to find a number that can be divided evenly into both the numerator and denominator of the fraction, you can continue to reduce the fraction. Ultimately, you will end up at the same result, $\frac{2}{3}$.

Reducing to Lowest Terms Problems

Reduce the following fractions to lowest terms.

1. $\frac{14}{21}$ 2. $\frac{66}{99}$

3. $\frac{20}{50}$ 4. $\frac{49}{84}$

5. $\frac{375}{525}$ 6. $\frac{24}{27}$

7. $\frac{42}{63}$ 8. $\frac{105}{165}$

9. $\frac{180}{260}$ 10. $\frac{68}{76}$

11. $\frac{18}{42}$ 12. $\frac{125}{600}$

13. List the following three fractions from highest to lowest.

$$\frac{5}{24}, \frac{12}{48}, \frac{9}{72}$$

14. Five relatives are to share in a $100,000 inheritance. According to the will, Rudi is to get $\frac{2}{16}$, Colleen is entitled to $\frac{7}{28}$, Bob gets $\frac{3}{9}$, Irene's share is $\frac{3}{36}$, and finally Brady gets $\frac{5}{24}$. Put the five relatives in order of highest share to lowest share.

Reducing to Lowest Terms Solutions

1. $\dfrac{14}{21} = \dfrac{14 \div 7}{21 \div 7} = \dfrac{2}{3}$

2. $\dfrac{66}{99} = \dfrac{66 \div 11}{99 \div 11} = \dfrac{6}{9} = \dfrac{6 \div 3}{9 \div 3} = \dfrac{2}{3}$

3. $\dfrac{20}{50} = \dfrac{20 \div 10}{50 \div 10} = \dfrac{2}{5}$

(Note: You can cancel zeros in the numerator and denominator as long as you cancel an equal number in both.)

$\dfrac{2\cancel{0}}{5\cancel{0}} = \dfrac{2}{5}$

4. $\dfrac{49}{84} = \dfrac{49 \div 7}{84 \div 7} = \dfrac{7}{12}$

5. $\dfrac{375}{525} = \dfrac{375 \div 25}{525 \div 25} = \dfrac{15}{21} = \dfrac{15 \div 3}{21 \div 3} = \dfrac{5}{7}$

Remember: How many quarters in $3.75? How many quarters in $5.25?

6. $\dfrac{24}{27} = \dfrac{24 \div 3}{27 \div 3} = \dfrac{8}{9}$

7. $\dfrac{42}{63} = \dfrac{42 \div 7}{63 \div 7} = \dfrac{6}{9} = \dfrac{6 \div 3}{9 \div 3} = \dfrac{2}{3}$

8. $\dfrac{105}{165} = \dfrac{105 \div 5}{165 \div 5} = \dfrac{21}{33} = \dfrac{21 \div 3}{33 \div 3} = \dfrac{7}{11}$

9. $\dfrac{180}{260} = \dfrac{18\cancel{0}}{26\cancel{0}} = \dfrac{18 \div 2}{26 \div 2} = \dfrac{9}{13}$

10. $\dfrac{68}{76} = \dfrac{68 \div 4}{76 \div 4} = \dfrac{17}{19}$

11. $\dfrac{18}{42} = \dfrac{18 \div 6}{42 \div 6} = \dfrac{3}{7}$

12. $\dfrac{125}{600} = \dfrac{125 \div 25}{600 \div 25} = \dfrac{5}{24}$

13. In a later section, we will discuss an easier solution path to this problem, but for now we must use the one skill we have developed, and that is, reducing each fraction to lowest terms.

There is no number that will divide evenly into both 5 and 24, so $\dfrac{5}{24}$ is already reduced to lowest terms. The largest number that will divide evenly into 12 and 48 is 12. $\dfrac{12}{48}$ reduces to $\dfrac{1}{4}$. Finally, the largest number that divides evenly into 9 and 72 is 9. $\dfrac{9}{72}$ reduces to $\dfrac{1}{8}$. So the three fractions are $\dfrac{5}{24}$, $\dfrac{1}{4}$, and $\dfrac{1}{8}$. It is easy to see that $\dfrac{1}{4}$ is larger than $\dfrac{1}{8}$, but where does $\dfrac{5}{24}$ fit into the picture? Well, $\dfrac{5}{24}$ is larger than $\dfrac{4}{24}$ but smaller than $\dfrac{6}{24}$. $\dfrac{4}{24}$ reduces to $\dfrac{1}{6}$ and $\dfrac{6}{24}$ reduces to $\dfrac{1}{4}$. So, $\dfrac{5}{24}$ is larger than $\dfrac{1}{6}$, but smaller than $\dfrac{1}{4}$. If it is larger than $\dfrac{1}{6}$, it is certainly larger than $\dfrac{1}{8}$.

$\dfrac{12}{48}$, $\left(\dfrac{1}{4}\right)$ is larger than $\dfrac{5}{24}$, and $\dfrac{5}{24}$ is larger than $\dfrac{9}{72}$ $\left(\dfrac{1}{8}\right)$. The correct sequence is $\dfrac{12}{48}$, $\dfrac{5}{24}$, and $\dfrac{9}{72}$.

The relationship would have been easier for us to see had we only reduced each fraction so that each denominator was 24. Dividing both 12 and 48 by 2, we would get $\dfrac{6}{24}$. Dividing both 9 and 72 by 3, we would get $\dfrac{3}{24}$. Now, it is easy to see that $\dfrac{6}{24}$ is larger than $\dfrac{5}{24}$ which, in turn, is larger than $\dfrac{3}{24}$. The number 24 is called the common denominator and is the key to that easier solution. As promised, we will study common denominators in a future section.

6

14. If you reduce each fraction to lowest terms the result would be:

$$\frac{2}{16} = \frac{1}{8}$$

$$\frac{3}{9} = \frac{1}{3}$$

$$\frac{7}{28} = \frac{1}{4}$$

$$\frac{3}{36} = \frac{1}{12}$$

$$\frac{5}{24} = \frac{5}{24}$$

The first four fractions are easy to sequence. If you keep the value of the numerator the same (in this case, 1) you decrease the value of the fraction as you increase the denominator. $\frac{1}{3}$ is larger than $\frac{1}{4}$, which is larger than $\frac{1}{8}$, which is larger than $\frac{1}{12}$. From the previous problem, we know that $\frac{5}{24}$ is smaller than $\frac{1}{4}$ but larger than $\frac{1}{8}$. Therefore, the proper sequence of fractions is:

$$\frac{3}{9} \; (\frac{1}{3}), \; \frac{7}{28} \; (\frac{1}{4}), \; \frac{5}{24}, \; \frac{2}{16} \; (\frac{1}{8}), \text{ and } \frac{3}{36} \; (\frac{1}{12})$$

The proper sequence of relatives is: Bob, Colleen, Brady, Rudi, and Irene.

Key Point

Fractional numbers that are greater than 1 can be written in two different forms. One form is much better to work with when you are multiplying or dividing fractions. The other is better when you have to judge relative size. So, it is important that you are able to work with both and change from one form to the other.

$\frac{19}{5}$ is an Improper Fraction.

An Improper Faction is a fraction with a numerator that is larger than the denominator.

As we will see later in this book, this is the form we will use when we are multiplying or dividing fractions. However, in this form, it is difficult to evaluate its relative size.

Is $\frac{19}{5}$ larger than 5?

Is $\frac{19}{5}$ smaller than 3?

Would you rather earn 4 times more than you do now or $\frac{19}{5}$ more than you do now?

In order to judge relative size, we actually perform the division that this fraction defines. We divide 19 by 5. 5 goes into 19, 3 times with 4 remainder. The remainder is written over the original denominator, in this example, 5, forming the fraction $\frac{4}{5}$. Then we write this fraction next to the whole number 3.

$$\frac{19}{5} = 3\frac{4}{5}$$

The final result is $3\frac{4}{5}$. This is called a **Mixed Number**. A mixed number is made up of 2 parts, a whole number and a fraction that is greater than zero but less than 1.

Key Point

To review, in order to change an improper fraction to a mixed number, divide the denominator of the improper fraction into the numerator. Write the remainder over the original denominator. This fractional remainder is written next to the whole number. Remember, if we start with a fraction of 5, we end with a fraction of 5.

$3\frac{4}{5}$ is a number that is greater than 3, but less than 4. So:

Is $\frac{19}{5}$ larger than 5? **NO**

Is $\frac{19}{5}$ smaller than 3? **NO**

Would you rather earn 4 times more than you do now or $\frac{19}{5}$ more than you do now? **4 times more.**

Let's try some more problems.

Example: Change $\frac{51}{8}$ to a mixed number.

Solution: $\frac{51}{8}$ $51 \div 8 = 6$ with a remainder of 3

   $\frac{51}{8} = 6\frac{3}{8}$

Example: Change $\frac{13}{5}$ to a mixed number.

Solution: $\frac{13}{5}$ $13 \div 5 = 2$ with a remainder of 3

   $\frac{13}{5} = 2\frac{3}{5}$

Changing From a Mixed Number to an Improper Fraction

A visual representation of $3\frac{4}{5}$ would be 3 rows, each with 5 blocks and a fourth row with only 4 blocks.

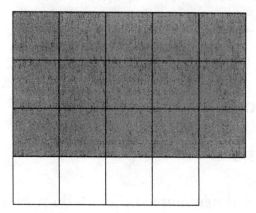

If we consider 5 blocks as a full row, then the last row is only $\frac{4}{5}$ of a row. The fastest way to count how many blocks there are is to consider that there are 3 full rows with 5 blocks each. To find the total number of blocks in these 3 rows, we would multiply 3 by 5, which is 15. To find the total number of blocks in the figure, we would have to add the 4 blocks from the partial row.

$$3 * 5 = 15$$

$$15 + 4 = 19$$

Key Point

To change a mixed number to an improper fraction, multiply the whole number by the denominator of the fraction and add this product to the numerator of the fraction. The sum is placed over the original denominator.

Changing $3\frac{4}{5}$ back to an improper fraction, we would multiply 3 by 5 (15) and add 4 = 19. 19 is placed over the original denominator (5). $3\frac{4}{5} = \frac{19}{5}$.

Regardless of which direction we move in, we start with a fraction of 5 and we end with a fraction of 5.

Example: Change $7\frac{4}{5}$ to an improper fraction.

Solution:

Step 1: Multiply the whole number (7) by the denominator (5).

$5 \times 7 = 35$

Step 2: Add the product (35) to the numerator (4).

$35 + 4 = 39$

Step 3: Put the result (39) over the original denominator (5).

$$\textbf{Mixed Number} \rightarrow 7\frac{4}{5} = \frac{39}{5} \leftarrow \textbf{Improper Fraction}$$

Example: Change $5\frac{3}{7}$ to an improper fraction.

Solution: $7 \times 5 = 35$ $35 + 3 = 38$

$$5\frac{3}{7} = \frac{38}{7}$$

Improper Fractions and Mixed Numbers Sample Problems

Directions: Change the following improper fractions to mixed numbers and reduce to lowest terms.

1. $\frac{15}{8}$ 2. $\frac{29}{12}$

3. $\frac{62}{11}$ 4. $\frac{110}{25}$

5. $\frac{42}{24}$ 6. $\frac{82}{9}$

11

Directions: Change the following mixed numbers to improper fractions.

7. $3\frac{4}{9}$ 8. $1\frac{1}{4}$

9. $11\frac{2}{3}$ 10. $7\frac{3}{4}$

11. $8\frac{6}{7}$ 12. $9\frac{2}{5}$

13. Of the following fractions, which is the largest?

$$\frac{7}{4}, \frac{3}{2}, 3\frac{1}{8}, \frac{10}{3}, \frac{15}{5}$$

14. Arrange the following fractions in descending order.

$$\frac{19}{6}, \frac{17}{3}, 4\frac{2}{5}, \frac{27}{10}, \frac{23}{4}$$

Improper Fractions and Mixed Numbers Sample Problem Solutions

1.   8 goes into 15, once with 7 remainder

$$\frac{15}{8} = 1\frac{7}{8}$$

2.   12 goes into 29, twice with 5 remainder

$$\frac{29}{12} = 2\frac{5}{12}$$

3.   11 goes into 62, 5 times with 7 remainder

$$\frac{62}{11} = 5\frac{7}{11}$$

4.   25 goes into 110, 4 times with 10 remainder

$$\frac{110}{25} = 4\frac{10}{25} = 4\frac{2}{5}$$

5.   24 goes into 42, once with 18 remainder

$$\frac{42}{24} = 1\frac{18}{24} = 1\frac{3}{4}$$

6. 9 goes into 82, 9 times with 1 remainder

$$\frac{82}{9} = 9\frac{1}{9}$$

7. 9 x 3 + 4 = 31

$$3\frac{4}{9} = \frac{31}{9}$$

8. 4 x 1 + 1 =

$$1\frac{1}{4} = \frac{5}{4}$$

9. 3 x 11 + 2 = 35

$$11\frac{2}{3} = \frac{35}{3}$$

10. 4 x 7 + 3 = 31

$$7\frac{3}{4} = \frac{31}{4}$$

11. 7 x 8 + 6 = 62

$$8\frac{6}{7} = \frac{62}{7}$$

12. 5 x 9 + 2 = 47

$$9\frac{2}{5} = \frac{47}{5}$$

13.

$$\frac{7}{4} = 1\frac{3}{4}$$

$$\frac{3}{2} = 1\frac{1}{2}$$

$$3\frac{1}{8} = 3\frac{1}{8}$$

$$\frac{10}{3} = 3\frac{1}{3}$$

$$\frac{15}{5} = 3$$

After all of the improper fractions have been converted to mixed numbers, it is easy to see that $\frac{10}{3}$ or $3\frac{1}{3}$ is the largest quantity.

14.
$$\frac{19}{6} = 3\frac{1}{6}$$

$$\frac{17}{3} = 5\frac{2}{3}$$

$$4\frac{2}{5} = 4\frac{2}{5}$$

$$\frac{27}{10} = 2\frac{7}{10}$$

$$\frac{23}{4} = 5\frac{3}{4}$$

The largest number is either $5\frac{2}{3}$ or $5\frac{3}{4}$. To be precise in our determination we would have to express both fractions in some common fashion. This is called a common denominator and is a topic for later discussion. For now, it is only necessary to know that each time you add 1 to both the numerator and denominator of a fraction, the resulting fraction is larger in size than the original fraction. Therefore, $\frac{2}{3}$ is less than $\frac{(2+1)}{(3+1)}$ or $\frac{3}{4}$. So, $5\frac{3}{4}$ is larger than $5\frac{2}{3}$. The remainder of the sequence follows without difficulty, as it is easy to see that $4\frac{2}{5}$ is larger than $3\frac{1}{6}$, which, in turn, is larger than $2\frac{7}{10}$. In their original form the sequence in descending order is:

$$\frac{23}{4}, \frac{17}{3}, 4\frac{2}{5}, \frac{19}{6}, \frac{27}{10}$$

Section 3: Adding Fractions with Common Denominators

Key Point

The denominator of a fraction **names** the fraction. You can only add or subtract fractions with the same name, or denominator. For example, you can add $\frac{3}{7} + \frac{2}{7} + \frac{1}{7}$ together because they are **all fractions of 7**. They have a **Common Denominator of 7**. When this is the case, you need only add the numerators together. **Do Not add the denominators** together. If you add fractions of 7 together, you end up with a fraction of 7. The name does not change.

So, $\frac{3}{7} + \frac{2}{7} + \frac{1}{7} = \frac{6}{7}$

Example: Find the sum of $\frac{4}{9}$ and $\frac{8}{9}$.

Solution: In this case the **common denominator is 9** – both numbers are fractions of 9. We add the numerators and the denominator remains the same.

$$\frac{4}{9} + \frac{8}{9} = \frac{12}{9}$$

We can **divide** numerator and denominator **by 3** and reduce the fraction to $\frac{4}{3}$.

Additionally, we can write the result as a mixed fraction $1\frac{1}{3}$.

Example: Add $\frac{5}{8} + \frac{5}{8}$ and reduce to lowest terms.

Solution: $\frac{5}{8} + \frac{5}{8} = \frac{5+5}{8} = \frac{10}{8}$

$\frac{10}{8} = \frac{10 \div 2}{8 \div 2} = \frac{5}{4}$

$\frac{5}{4} = 5 \div 4 = 1\frac{1}{4}$

Adding Fractions With a Common Denominator Problems

Directions: Convert to improper fractions if possible. Always reduce to lowest terms.

1. $\dfrac{6}{17} + \dfrac{3}{17} =$

2. $\dfrac{4}{9} + \dfrac{2}{9} + \dfrac{7}{9} =$

3. $\dfrac{9}{25} + \dfrac{11}{25} =$

4. $\dfrac{11}{40} + \dfrac{3}{40} + \dfrac{19}{40} =$

5. $\dfrac{3}{8} + \dfrac{5}{8} + \dfrac{7}{8} =$

6. $\dfrac{5}{12} + \dfrac{11}{12} =$

7. Each of the objects below has been divided so that all of its parts are of equal size. Express each shaded region as a fraction and find the sum of all the shaded areas.

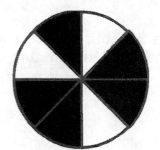  

Adding Fractions With a Common Denominator Solutions

1. $\dfrac{6}{17} + \dfrac{3}{17} = \dfrac{6+3}{17} = \dfrac{9}{17}$

2. $\dfrac{4}{9} + \dfrac{2}{9} + \dfrac{7}{9} = \dfrac{4+2+7}{9} = \dfrac{13}{9} = 1\dfrac{4}{9}$

3. $\dfrac{9}{25} + \dfrac{11}{25} = \dfrac{9+11}{25} = \dfrac{20}{25} = \dfrac{4}{5}$

4. $\dfrac{11}{40} + \dfrac{3}{40} + \dfrac{19}{40} = \dfrac{11+3+19}{40} = \dfrac{33}{40}$

5. $\dfrac{3}{8} + \dfrac{5}{8} + \dfrac{7}{8} = \dfrac{3+5+7}{8} = \dfrac{15}{8} = 1\dfrac{7}{8}$

6. $\dfrac{5}{12} + \dfrac{11}{12} = \dfrac{5+11}{12} = \dfrac{16}{12} = \dfrac{4}{3} = 1\dfrac{1}{3}$

7.   In the first figure, which is a circle, five of the eight sections are shaded. The first fraction is therefore equal to $\frac{5}{8}$.

In the second figure, which is a rectangle, three of the eight sections are shaded. The third fraction is therefore equal to $\frac{3}{8}$.

Finally, in the third figure, which is a square, four of the eight sections are shaded. The third fraction is therefore equal to $\frac{4}{8}$.

Sum of the shaded areas $= \frac{5}{8} + \frac{3}{8} + \frac{4}{8} =$

Sum of the shaded areas $= \frac{5+3+4}{8} = \frac{12}{8} = 1\frac{1}{2}$

Section 4: Subtracting Fractions with Common Denominators

As in addition, you can only subtract fractions that have the same denominator (name). If they do, then solving this problem is just a matter of subtracting the numerators and writing the difference over the common denominator.

Example: Subtract $\dfrac{13}{30}$ from $\dfrac{19}{30}$.

Solution: Subtract 13 from 19, and place the difference, 6, over the common denominator of 30.

The number 6 divides evenly into both the numerator and denominator. So, $\dfrac{6}{30}$ can be reduced to $\dfrac{1}{5}$.

Example: Of a class of 25 students, $\dfrac{2}{5}$ are out with the flu.   Express, as a fraction, the number of students left in class.

Solution: 25 Total Students

Flu	Flu			

$\dfrac{2}{5}$ Have the Flu.

To answer the question, it is not important to know the total number of students. $\dfrac{2}{5}$ represents the fraction of students who have the flu. All of the students would be represented by the number 1. To find the fraction that represents the students still in the classroom, we subtract $\frac{2}{5}$ from 1. How many fifths is the number 1? It's all 5 of the fifths. We can write the number 1 as $\frac{5}{5}$, and the subtraction as:

$$\frac{5}{5} - \frac{2}{5} = \frac{3}{5}$$

Subtracting Fractions With a Common Denominator Problems

1. $\frac{5}{9} - \frac{2}{9} =$ 2. $\frac{7}{12} - \frac{1}{12} =$

3. $\frac{23}{24} - \frac{8}{24} =$ 4. $\frac{9}{40} - \frac{5}{40} =$

5. $\frac{8}{11} - \frac{3}{11} =$ 6. $\frac{19}{30} - \frac{13}{30} =$

7. **Of a class of 24 students, half are out with the flu and 6 more are at band practice. Express as a fraction the number of students left in the class.**

Subtracting Fractions With a Common Denominator Solution

1. $\frac{5}{9} - \frac{2}{9} = \frac{5-2}{9} = \frac{3}{9} = \frac{1}{3}$

2. $\frac{7}{12} - \frac{1}{12} = \frac{7-1}{12} = \frac{6}{12} = \frac{1}{2}$

3. $\frac{23}{24} - \frac{8}{24} = \frac{23-8}{24} = \frac{15}{24} = \frac{5}{8}$

4. $\frac{9}{40} - \frac{5}{40} = \frac{9-5}{40} = \frac{4}{40} = \frac{1}{10}$

5. $\frac{8}{11} - \frac{3}{11} = \frac{8-3}{11} = \frac{5}{11}$

6. $\frac{19}{30} - \frac{13}{30} = \frac{19-13}{30} = \frac{6}{30} = \frac{1}{5}$

7. Half of the class is out with the flu. That equals 12 of the 24 students or $\frac{12}{24}$.

Six students are at band practice or $\frac{6}{24}$.

The fraction of the students remaining is the difference between a full class $\frac{24}{24}$ and those absent $\frac{18}{24}$.

Fraction present = $\frac{24}{24} - \frac{18}{24} = \frac{(24-18)}{24} = \frac{6}{24}$ or $\frac{1}{4}$

Section 5: Finding and Using a Common Denominator

Key Point

If the fractions you wish to add or subtract do not have a common name (common denominator), such as $\frac{3}{7}$ and $\frac{2}{5}$, then you have to rename each one so that they do both have the same denominator.

We know that $\frac{3}{7} = \frac{6}{14} = \frac{9}{21} = \frac{12}{28} = \frac{15}{35}$

We know that $\frac{2}{5} = \frac{4}{10} = \frac{6}{15} = \frac{8}{20} = \frac{10}{25} = \frac{12}{30} = \frac{14}{35}$

We can write both fractions, $\frac{3}{7}$ and $\frac{2}{5}$, as fractions of 35. 35 is the **Common Denominator.**

Key Point

For the two fractions that we are working with, the common denominator is a number that both 7 and 5 can divide evenly into. If you cannot immediately think of such a number, the product of the 2 denominators will always work. This may not get you the smallest common denominator, but it will give you a common denominator.

In our example, the product of the two denominators is 35.

$\frac{3}{7}$ ↘ $\frac{?}{35}$ 7 goes into 35, 5 times

     5 times 3 is 15.

$\frac{3}{7} = \frac{15}{35}$

$\frac{2}{5}$ ↘ $\frac{?}{35}$ 5 goes into 35, 7 times

     7 times 2 is 14.

$\frac{2}{5} = \frac{14}{35}$

Now we can add the fractions together.

$$\frac{3}{7} + \frac{2}{5} = \frac{15}{35} + \frac{14}{35} = \frac{29}{35}$$

Example: Subtract $\frac{2}{9}$ from $\frac{5}{6}$.

21

Solution: First we must find a common denominator – a number that both 6 and 9 divide evenly into. If we have trouble finding one, we can always use their product, 54. But this is not the smallest common denominator. The smallest number that both 6 and 9 divide evenly into is 18.

6 goes into 18, 3 times

9 goes into 18, 2 times

$$\frac{5}{6} \rightarrow \frac{?}{18}$$ 6 goes into 18, 3 times

3 times 5 is 15.

$$\frac{5}{6} = \frac{15}{18}$$

$$\frac{2}{9} \rightarrow \frac{?}{18}$$ 9 goes into 18, 2 times

2 times 2 is 4.

$$\frac{2}{9} = \frac{4}{18}$$

$$\frac{5}{6} - \frac{2}{9} = \frac{15}{18} - \frac{4}{18} = \frac{11}{18}$$

Finding a Common Denominator Problems

1.   Change $\frac{4}{5}$ to a fraction of 20.

2.   Change $\frac{7}{10}$ to a fraction of 20.

3.   Change $\frac{3}{4}$ to a fraction of 12.

4.   Change $\frac{5}{6}$ to a fraction of 12.

Finding a Common Denominator Solutions

1. $\frac{4}{5} = \frac{16}{20}$ 5 goes into 20, 4 times; 4 times 4 is 16.

$$\frac{4}{5} \rightarrow \frac{16}{20}$$

2. $\frac{7}{10} = \frac{14}{20}$ 10 goes into 20, 2 times; 2 times 7 is 14.

$$\frac{7}{10} \rightarrow \frac{14}{20}$$

3. $\dfrac{3}{4} = \dfrac{9}{12}$ 4 goes into 12, 3 times; 3 times 3 is 9.

$$\dfrac{2}{4} \rightarrow \dfrac{9}{12}$$

4. $\dfrac{5}{6} = \dfrac{10}{12}$ 6 goes into 12, 2 times; 2 times 5 is 10.

$$\dfrac{5}{6} \rightarrow \dfrac{10}{12}$$

Adding and Subtracting Fractions With Unlike Denominators Problems

DIRECTIONS: Solve the following problems. Answers should be in simplified form.

1. Add $\dfrac{2}{3}$ and $\dfrac{1}{4}$.

2. $\dfrac{2}{9} + \dfrac{1}{6} =$

3. $\dfrac{5}{12} + \dfrac{2}{3} + \dfrac{7}{24} =$

4. Add $\dfrac{6}{7}$ and $\dfrac{4}{21}$.

5. Add $\dfrac{4}{5}$ and $\dfrac{7}{10}$.

6. $\dfrac{5}{6} + \dfrac{3}{4} =$

7. $\dfrac{3}{4} - \dfrac{2}{3} =$

8. $\dfrac{1}{2} - \dfrac{5}{24} =$

9. $\dfrac{11}{12} - \dfrac{5}{18} =$

10. Subtract $\dfrac{2}{5}$ from $\dfrac{21}{25}$.

11. **Before the football game started, Richard and 23 of his closest friends decided to order pizza. When asked their preference, half said anything's okay, $\frac{1}{6}$ said anything but anchovies, and another quarter requested sausage and peppers. The remainder wanted nothing extra. What fraction of Richard's friends like plain pizza?**

12. **Colleen spends $\frac{1}{4}$ of her monthly take home pay on rent, $\frac{1}{8}$ on food and clothes, $\frac{3}{40}$ on heat and electricity, $\frac{1}{5}$ on insurance and repayment of loans, $\frac{1}{10}$ on transportation, and $\frac{1}{16}$ on phone, water, and other bills and miscellaneous household items. What fraction of her take home pay is left for entertainment and savings?**

Adding and Subtracting Fractions With Unlike Denominators Solutions

1. The common denominator is 12.

$$\frac{2}{3} \rightarrow \frac{8}{12} \qquad \frac{1}{4} \rightarrow \frac{3}{12}$$

$$\frac{2}{3} + \frac{1}{4} = \frac{8}{12} + \frac{3}{12} = \frac{11}{12}$$

2. The common denominator is 18.

$$\frac{2}{9} \rightarrow \frac{4}{18} \qquad \frac{1}{6} \rightarrow \frac{3}{18}$$

$$\frac{2}{9} + \frac{1}{6} = \frac{4}{18} + \frac{3}{18} = \frac{7}{18}$$

3. The common denominator is 24.

$$\frac{5}{12} \rightarrow \frac{10}{24} \qquad \frac{2}{3} \rightarrow \frac{16}{24}$$

$$\frac{5}{12} + \frac{2}{3} + \frac{7}{24} = \frac{10}{24} + \frac{16}{24} + \frac{7}{24} = \frac{33}{24} = 1\frac{9}{24} = 1\frac{3}{8}$$

4. The common denominator is 21.

$$\frac{6}{7} \rightarrow \frac{18}{21}$$

$$\frac{6}{7} + \frac{4}{21} = \frac{18}{21} + \frac{4}{21} = \frac{22}{21} = 1\frac{1}{21}$$

5. The common denominator is 10.

$$\frac{4}{5} \rightarrow \frac{8}{10}$$

$$\frac{4}{5} + \frac{7}{10} = \frac{8}{10} + \frac{7}{10} = \frac{15}{10} = 1\frac{5}{10} = 1\frac{1}{2}$$

6. The common denominator is 12.

$$\frac{3}{4} \rightarrow \frac{9}{12} \qquad \frac{5}{6} \rightarrow \frac{10}{12}$$

$$\frac{3}{4} + \frac{5}{6} = \frac{9}{12} + \frac{10}{12} = \frac{19}{12} = 1\frac{7}{12}$$

7. The common denominator is 12.

$$\frac{3}{4} \rightarrow \frac{9}{12} \qquad \frac{2}{3} \rightarrow \frac{8}{12}$$

$$\frac{3}{4} - \frac{2}{3} = \frac{9}{12} - \frac{8}{12} = \frac{1}{12}$$

8. The common denominator is 24.

$$\frac{1}{2} \rightarrow \frac{12}{24}$$

$$\frac{1}{2} - \frac{5}{24} = \frac{12}{24} - \frac{5}{24} = \frac{7}{24}$$

9. The common denominator is 36.

$$\frac{11}{12} \rightarrow \frac{33}{36} \qquad \frac{5}{18} \rightarrow \frac{10}{36}$$

$$\frac{11}{12} - \frac{5}{18} = \frac{33}{36} - \frac{10}{36} = \frac{23}{36}$$

10. The common denominator is 25.

$$\frac{2}{5} \rightarrow \frac{10}{25}$$

$$\frac{21}{25} - \frac{2}{5} = \frac{21}{25} - \frac{10}{25} = \frac{11}{25}$$

11. First, we want to add the fractions $\frac{1}{2}$, $\frac{1}{6}$, and $\frac{1}{4}$ together. We could get a common denominator by multiplying 2, 4, and 6 together, but this is not the smallest common denominator. Although you could use 24 to solve the problem, 12 is the smallest common denominator.

2 goes into 12, 6 times so $\frac{1}{2} = \frac{6}{12}$

6 goes into 12, 2 times so $\dfrac{1}{6} = \dfrac{2}{12}$

4 goes into 12, 3 times so $\dfrac{1}{4} = \dfrac{3}{12}$

The sum of these three fractions is $\dfrac{11}{12}$. The fraction of friends who prefer their pizza plain is equal to:

$$\dfrac{12}{12} - \dfrac{11}{12} = \dfrac{1}{12}$$

12. First, we have to find the common denominator of 4, 5, 8, 10, 16, 40. 4, 5, 8 and 10 all divide evenly into 40. 16, however, does not. The next higher multiple of 40 is 80. We know that 4, 5, 8, and 10 will divide evenly into 80 because they divided evenly into 40. 80 is also divisible by 16, so 80 is our common denominator.

4 goes into 80, 20 times; so $\dfrac{1}{4} = \dfrac{20}{80}$

8 goes into 80, 10 times; so $\dfrac{1}{8} = \dfrac{10}{80}$

40 goes into 80, 2 times; so $\dfrac{3}{40} = \dfrac{6}{80}$

5 goes into 80, 16 times; so $\dfrac{1}{5} = \dfrac{16}{80}$

10 goes into 80, 8 times; so $\dfrac{1}{10} = \dfrac{8}{80}$

16 goes into 80, 5 times; so $\dfrac{1}{16} = \dfrac{5}{80}$

Now add the fractions together to get a total.

$$\dfrac{20}{80} + \dfrac{10}{80} + \dfrac{6}{80} + \dfrac{16}{80} + \dfrac{8}{80} + \dfrac{5}{80} =$$

The sum of these six fractions is $\dfrac{65}{80}$. The amount of income remaining for entertainment and savings is:

$$\dfrac{80}{80} - \dfrac{65}{80} = \dfrac{15}{80} \text{ or } \dfrac{3}{16}$$

Colleen has $\dfrac{3}{16}$ of her take home pay left for entertainment and savings.

Key Point

When adding or subtracting mixed numbers you will perform the appropriate operation on the whole numbers separately from the fractions. Actually, it is as if you were seeking the solution to **two** problems instead of **one**. The fraction in your final answer, of course, will be simplified (reduced to lowest terms), if possible.

Section 6: Adding and Subtracting Mixed Numbers

Let's try some examples.

Example: Add $3 \frac{5}{18}$ and $7 \frac{1}{18}$

Solution: $3 \frac{5}{18}$ $3 + 7 = 10$ addition of the whole numbers

$+7 \frac{1}{18}$ $\frac{5}{18} + \frac{1}{18} = \frac{6}{18}$ addition of the fractions

$+10 \frac{6}{18} = 10 \frac{1}{3}$ combine the two and reduce the fraction to

lowest terms

Example: The difference between $9 \frac{3}{8}$ and $2 \frac{1}{4}$ is:

1) $11 \frac{1}{4}$

2) $5 \frac{1}{4}$

3) $7 \frac{5}{8}$

4) $7 \frac{1}{8}$

5) $11 \frac{4}{12}$

Solution: The word **difference** is your clue that this is a subtraction problem.

$$9 \frac{3}{8}$$
$$-2 \frac{1}{4}$$
$$\overline{7 \frac{1}{8}}$$

Step 1: Convert the fractions so that they have a common denominator then subtract.

$$\frac{3}{8} - \frac{1}{4} = \frac{3}{8} - \frac{2}{8} = \frac{1}{8}$$

Step 2: Subtract the whole number portion of the problem.

$$9 - 2 = 7$$

Step 3: Put the parts together.

$$7 \frac{1}{8}$$

Choice "4" is the correct response.

HINT: It is helpful in a problem of this type to first determine the approximate size of the answer. You are asked to find the difference between a number larger than 9 and a number slightly larger than 2. You will get an answer slightly larger than 9 minus 2, or 7. This eliminates responses 1, 2, and 5. Careful inspection would even eliminate response 3. This can save you a great deal of time.

Key Point

If the problem involves subtracting a larger fraction from a smaller one, you will have to "borrow" from the whole number.

Example: Subtract $5 \frac{7}{8}$ from $11 \frac{3}{8}$.

Solution: $\frac{7}{8}$ is larger than $\frac{3}{8}$. How can you solve this problem?

You will "borrow" 1 from the whole number 11, decreasing the 11 in size by 1 and leaving 10. The one you "borrowed" is added to the $\frac{3}{8}$ in the form of $\frac{8}{8}$.

$$\frac{3}{8} + \frac{8}{8} = \frac{11}{8}$$

So, you have transformed $11\,\frac{3}{8}$ into its equivalent form $10\,\frac{11}{8}$. Now you can perform the subtraction. Remember to leave your answer in simplified form.

$$11\,\frac{3}{8} = 10\,\frac{11}{8}$$
$$-\ 5\,\frac{7}{8} = \ 5\,\frac{7}{8}$$
$$\overline{\qquad\qquad\qquad 5\,\frac{4}{8} = 5\,\frac{1}{2}}$$

Step 1: Convert the fractions so that they have a common denominator and that the number being subtracted is smaller than the number it is subtracted from. Then subtract.

$$\frac{3}{8} \text{ is changed to } \frac{11}{8}, \frac{11}{8} - \frac{7}{8} = \frac{4}{8} = \frac{1}{2}$$

Step 2: Subtract the whole number portion of the problem.

$$10 - 5 = 5$$

Step 3: Put the parts together.

$$5\,\frac{1}{2}$$

Key Point

IN REVIEW:
To perform the operation of "borrowing", remember the following rules:
I.   Decrease the whole number by one.
II.   Add the denominator to the numerator.
III.   Put new numerator over the original denominator.

Example: $11\,\frac{3}{8} = 10\,\frac{3+8}{8} = 10\,\frac{11}{8}$

Directions: Perform the appropriate operation.

1. $3\frac{5}{12} + 6\frac{2}{12}$ 2. $5\frac{4}{9} + 12\frac{2}{9}$

3. $1\frac{1}{7} + 4\frac{3}{7} + 2\frac{5}{7}$ 4. $20\frac{1}{2} - 10\frac{5}{16}$

5. $9\frac{4}{15} - 7\frac{7}{15}$

6. The length of the wall in Bob's bedroom is $12\frac{1}{2}$ feet. Along this wall he plans to put a dresser measuring 3 feet 11 inches and a desk to hold his computer and printer. The desk measures 6 feet 5 inches. What is the maximum distance these two pieces of furniture can be apart?

Working With Mixed Numbers Solutions

1. $\begin{array}{r} 3\frac{5}{12} \\ +6\frac{2}{12} \\ \hline 9\frac{7}{12} \end{array}$

$3 + 6 = 9$

$\frac{5}{12} + \frac{2}{12} = \frac{7}{12}$

This fraction cannot be reduced any lower.

2. $\begin{array}{r} 5\frac{4}{9} \\ +12\frac{2}{9} \\ \hline 17\frac{6}{9} = 17 \end{array}$

$5 + 12 = 17$

$\frac{4}{9} + \frac{2}{9} = \frac{6}{9}$

$\frac{2}{3}$ This fraction can be reduced by dividing by 3.

3. $\begin{array}{r} 1\frac{1}{7} \\ 4\frac{3}{7} \\ +2\frac{5}{7} \\ \hline 7\frac{9}{7} = 8\frac{2}{7} \end{array}$

$1 + 4 + 2 = 7$

$\frac{1}{7} + \frac{3}{7} + \frac{5}{7} = \frac{9}{7}$

The improper fraction in the answer must be reduced to a mixed number.

4.

$$20\frac{1}{2} = 20\frac{8}{16} \qquad 20 - 10 = 10$$

$$-10\frac{5}{16} = 10\frac{5}{16} \qquad \frac{8}{16} - \frac{5}{16} = \frac{3}{16}$$

$$10\frac{3}{16} \qquad \text{This fraction cannot be reduced.}$$

NOTE: In this problem, it was necessary to first find a common denominator (16) before you could solve the problem. Each skill you learn is used to help you perform other operations. It is important that you master each one before you go on to the next.

5.

$$9\frac{4}{15} = 8\frac{19}{15} \qquad 8 - 7 = 1$$

$$-7\frac{7}{15} = 7\frac{7}{15} \qquad \frac{19}{15} - \frac{7}{15} = \frac{12}{15}$$

$$1\frac{12}{15} = 1\frac{4}{5} \qquad \begin{array}{l}\text{It is necessary to reduce this fraction} \\ \text{to lowest terms by dividing by 3.}\end{array}$$

6. This problem involves several steps.

First, we have to determine how much space the furniture will take up.

There are 12 inches in a foot, so the dresser is $3\frac{11}{12}$ and the desk is

$6\frac{5}{12}$ feet. Adding these two figures together we get:

$$3\frac{11}{12} \qquad 3 + 6 = 9$$

$$+6\frac{5}{12} \qquad \frac{11}{12} + \frac{5}{12} = \frac{16}{12}$$

$$9\frac{16}{12} = 10\frac{4}{12} \qquad \begin{array}{l}\text{Don't reduce this fraction yet because} \\ \text{you will want your final answer to be} \\ \text{in inches or feet.}\end{array}$$

Now that we know the amount of space the two pieces take up, we can find the amount of space that would be left.

The maximum space between these two pieces of furniture would be the difference between $12\frac{6}{12}$ (the length of the wall) and $10\frac{4}{12}$ (the amount of space the two pieces would need).

$$12 \frac{6}{12}$$
$$+ \; 10 \; \frac{4}{12}$$
$$\overline{\quad 2 \; \frac{2}{12}}$$

$12 - 10 = 2$

$\frac{6}{12} + \frac{4}{12} = \frac{2}{12}$

$\frac{2}{12}$ feet equals 2 inches so the final answer is 2 feet 2 inches.

Section 7: Multiplying Fractions

1	2	3	4				
5	6	7	8				
9	10	11	12				

The figure above is divided into 24 equal parts.

24   Total Squares

16   Shaded Squares

12   Numbered Squares

$$\frac{16}{24} = \frac{2}{3} \text{ of the squares are shaded}$$

$$\frac{12}{16} = \frac{3}{4} \text{ of the shaded squares are numbered}$$

$$\frac{12}{24} = \frac{1}{2} \text{ of the squares are numbered}$$

Two-thirds of the parts are shaded. Three-quarters of the shaded parts are numbered. Put another way, $\frac{3}{4}$ of the $\frac{2}{3}$ are numbered. In an arithmetic problem, the word "of" translates into multiplication. Three-quarters of two-thirds equals $\frac{3}{4}$ times $\frac{2}{3}$.

We can see from the figure that 12 of the 24 squares are numbered, so we know that the result of the multiplication is $\frac{12}{24}$ or $\frac{1}{2}$. But what are the rules for multiplying two or more fractions together?

Key Point

Rule 1:	You do not need a common denominator in order to multiply fractions.
Rule 2:	When multiplying two or more fractions together, first rewrite any mixed numbers as improper fractions.
Rule 3:	Multiply the numerators together and place this product over the product of the denominators.

33

So, in our example:

$$\frac{3}{4} \times \frac{2}{3} = \frac{6}{12} = \frac{1}{2}$$

Example: Multiply $\frac{3}{5}$ by $\frac{2}{7}$.

Solution: $\frac{3}{5} \times \frac{2}{7} =$

$$\frac{3 \times 2}{5 \times 7} = \frac{6}{35}$$

Example: Solve $1\frac{1}{3}$ times $\frac{4}{5}$.

Solution: $1\frac{1}{3} \times \frac{4}{5} =$

Remember to convert $1\frac{1}{3}$ to an improper fraction.

$$1\frac{1}{3} = \frac{4}{3}$$

THEREFORE: $\frac{4}{3} \times \frac{4}{5} =$

$$\frac{4 \times 4}{3 \times 5} = \frac{16}{15} = 1\frac{1}{16}$$

Key Point

A procedure called **canceling** might make the operation of multiplication easier for you. This aid is similar to simplifying fractions, except that in canceling you are looking for a number that will divide evenly into any one of the numerators and any one of the denominators. For instance, in our example where we were multiplying $\frac{3}{4}$ and $\frac{2}{3}$, we could divide the numerator of the first fraction and the denominator of the second fraction by 3. Also, the denominator of the first fraction and the numerator of the second fraction are both divisible by 2.

$$\frac{\overset{1}{\cancel{3}}}{\underset{2}{\cancel{4}}} \times \frac{\overset{1}{\cancel{2}}}{\underset{1}{\cancel{3}}} = \frac{1 \times 1}{2 \times 1} = \frac{1}{2}$$

Key Point

> Cancelling makes the problem easier to solve. However, if you do not cancel, you will still get the correct answer.

Example: Multiply $\dfrac{4}{13}$ by $\dfrac{7}{8}$.

Solution: The number 4 divides evenly into 4 and into 8.

$$\dfrac{^{1}\cancel{4}}{13} \times \dfrac{7}{\cancel{8}_2}$$

4 goes into 4, 1 time

4 goes into 8, 2 times

The problem simplifies to:

$$\dfrac{1}{13} \times \dfrac{7}{2} = \dfrac{1 \times 7}{13 \times 2} = \dfrac{7}{26}$$

You could also solve this problem by first multiplying and then reducing the answer. However, the multiplications are larger and more time consuming. Also simplification is not as easy to see as the canceling was.

$$\dfrac{4}{13} \times \dfrac{7}{8} = \dfrac{4 \times 7}{13 \times 8} = \dfrac{28}{104} = \dfrac{28 \div 4}{104 \div 4} = \dfrac{7}{26}$$

Example: Solve $3\,\dfrac{1}{3} \times \dfrac{3}{20} \times \dfrac{2}{7}$

Solution: $3\,\dfrac{1}{3} \times \dfrac{3}{20} \times \dfrac{2}{7}$ **FIRST**, rewrite $3\,\dfrac{1}{3}$ as $\dfrac{10}{3}$.

$$\dfrac{10}{3} \times \dfrac{3}{20} \times \dfrac{2}{7}$$ **NEXT**, cancel where possible

10 goes into 10 and 20

3 goes into 3 and 3

2 goes into 2 and 2

$$\dfrac{^{1}\cancel{10}}{_{1}\cancel{3}} \times \dfrac{^{1}\cancel{3}}{_{1}\cancel{2}\,20} \times \dfrac{^{1}\cancel{2}}{7} = \dfrac{1}{7}$$

Now, try some practice problems.

Multiplication of Fractions Problems

DIRECTIONS: Perform the following multiplications and leave your answer in simplified form. (Watch for the different signs that indicate multiplication.)

1. $\dfrac{6}{11} \times \dfrac{2}{3}$

2. $4\dfrac{1}{2} \times \dfrac{4}{15}$

3. $\dfrac{2}{5} \cdot \dfrac{25}{8}$

4. $\dfrac{(7)}{(9)} \dfrac{(3)}{(28)}$

5. $2\dfrac{1}{4} \times \dfrac{16}{27}$

6. $\dfrac{12}{21} \times 2\dfrac{1}{9}$

7. Before the football game started, Richard and 23 of his friends decided to order pizza. When asked their preference, half said anything is okay, $\dfrac{1}{8}$ said anything but anchovies, and another quarter requested sausage and peppers. The remainder wanted nothing extra. What fraction of Richard's friends like plain pizza? You solved this problem in an earlier section and obtained the answer of $\dfrac{1}{12}$ of Richard's friends like plain pizza. Now determine, how many people comprise $\dfrac{1}{12}$ of the group.

8. One-tenth of Cherie's monthly budget goes for miscellaneous bills other than rent, heat and electric. Her phone bill represents one-quarter of the miscellaneous bills. What fraction of her total monthly budget is her phone allowance? If her total budget is $2,000, how much is her phone allowance?

Questions 9 and 10 refer to the following information. Use the grid at the end of each problem to record your answer.

Anneliese has picked out the perfect color of pink to paint her bedroom. The manager of the paint store says that to make that shade he needs to mix a $1\frac{1}{2}$ gallons of white paint with 3 quarts of red paint. Anneliese only needs half this amount.

9.   What fraction of a gallon of white paint should the manager use?

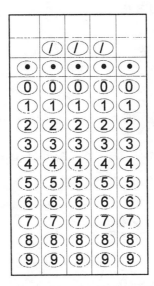

10.   What fraction of a gallon of red paint should the manager use?

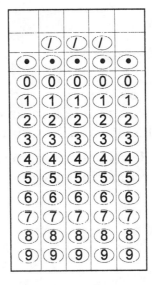

Multiplication of Fraction Solutions

1. $\dfrac{6}{11} \times \dfrac{2}{3} = \dfrac{6 \times 2}{11 \times 3} = \dfrac{12}{33} = \dfrac{4}{11}$ or $\dfrac{2\cancel{6}}{11} \times \dfrac{2}{\cancel{3}_1} = \dfrac{4}{11}$

2. $4\dfrac{1}{2} \times \dfrac{4}{15} = \dfrac{9}{2} \times \dfrac{4}{15} = \dfrac{^3\cancel{9}}{_1\cancel{2}} \times \dfrac{^2\cancel{4}}{_5\cancel{15}} = \dfrac{3 \times 2}{1 \times 5} = \dfrac{6}{5} = 1\dfrac{1}{5}$

3. $\dfrac{2}{5} \cdot \dfrac{25}{8} = \dfrac{^1\cancel{2}}{_1\cancel{5}} \cdot \dfrac{^5\cancel{25}}{_4\cancel{8}} = \dfrac{1 \times 5}{1 \times 4} = \dfrac{5}{4} = 1\dfrac{1}{4}$

4. $\dfrac{(7)}{(9)} \dfrac{(3)}{(28)} = \dfrac{^1\cancel{7}}{_3\cancel{9}} \times \dfrac{^1\cancel{3}}{_4\cancel{28}} = \dfrac{1 \times 1}{3 \times 4} = \dfrac{1}{12}$

5. $2\dfrac{1}{4} \times \dfrac{16}{27} = \dfrac{9}{4} \times \dfrac{16}{27} = \dfrac{^1\cancel{9}}{_1\cancel{4}} \times \dfrac{^4\cancel{16}}{_3\cancel{27}} = \dfrac{1 \times 4}{1 \times 3} = \dfrac{4}{3} = 1\dfrac{1}{3}$

6. $\dfrac{12}{21} \times 2\dfrac{1}{9} = \dfrac{^4\cancel{12}}{21} \times \dfrac{19}{_3\cancel{9}} = \dfrac{76}{63} = 1\dfrac{13}{63}$

7. You have already worked the problem to the point where you know that $\dfrac{1}{12}$ of the 24 people like plain pizza.

   $$\dfrac{1}{12} \text{ of } 24 = \dfrac{1}{_1\cancel{12}} \times \dfrac{^2\cancel{24}}{1} = 2$$

8. Cherie's phone allowance is $\dfrac{1}{4}$ of $\dfrac{1}{10}$.

   $$\dfrac{1}{4} \times \dfrac{1}{10} = \dfrac{1}{40}$$

   $$\dfrac{1}{_1\cancel{40}} \times \dfrac{^{50}\cancel{2000}}{1} = \$50$$

9. This problem, as well as the next, not only tests your knowledge of fractions, it also gives you practice in answering a problem using one of the alternative forms that the GED exam provides.

   To get the correct shade, but half the amount, they have to use half of each color. To find the amount of white paint to use, they have to take half of $1\dfrac{1}{2}$.

   $$\dfrac{1}{2} \times 1\dfrac{1}{2} = \dfrac{1}{2} \times \dfrac{3}{2} = \dfrac{3}{4}$$

They will use $\frac{3}{4}$ of a gallon of white paint.

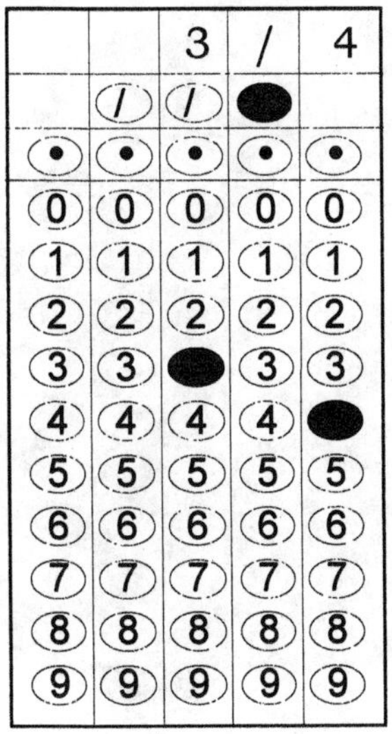

or

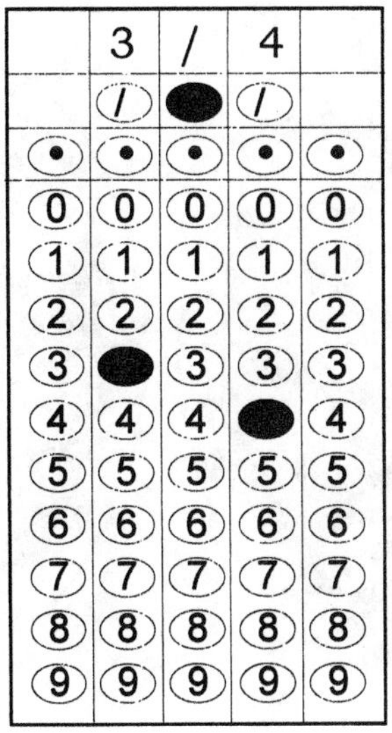

or

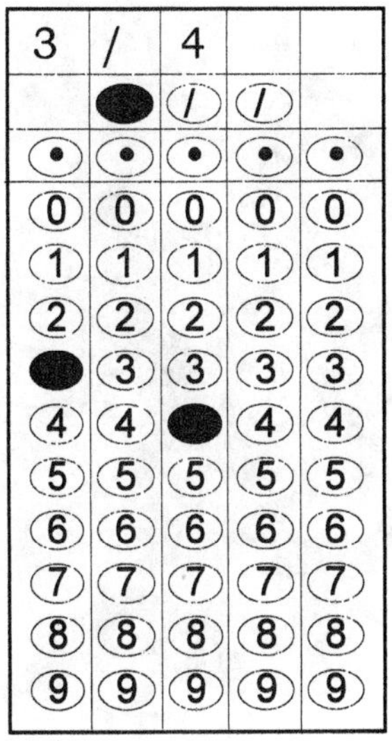

10. Since the original formula called for 3 gallons of red paint, they will use half of 3.

$$\frac{1}{2} \times 3 = \frac{3}{2}$$

They will use $\frac{3}{2}$ of a gallon of red paint.

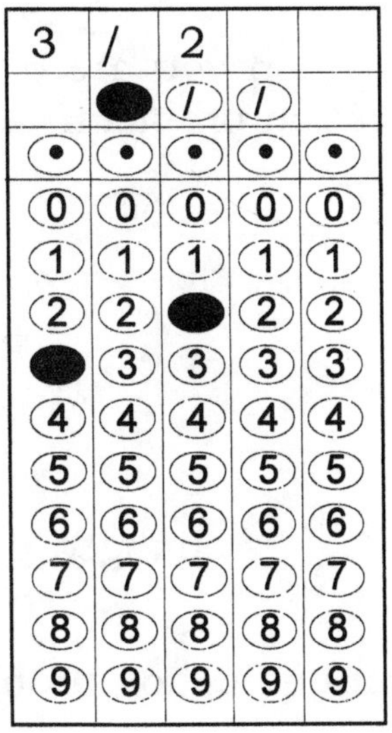

or

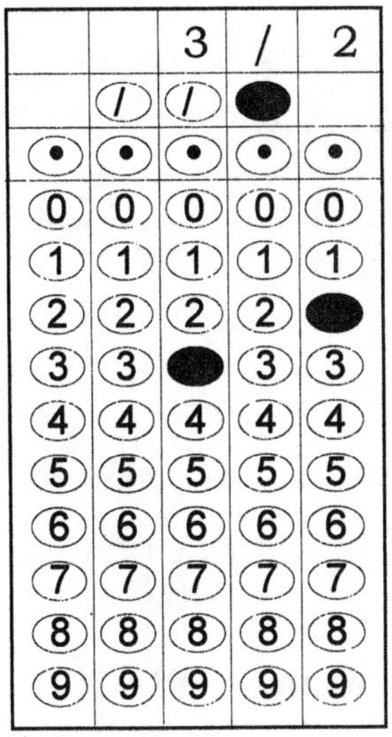

or

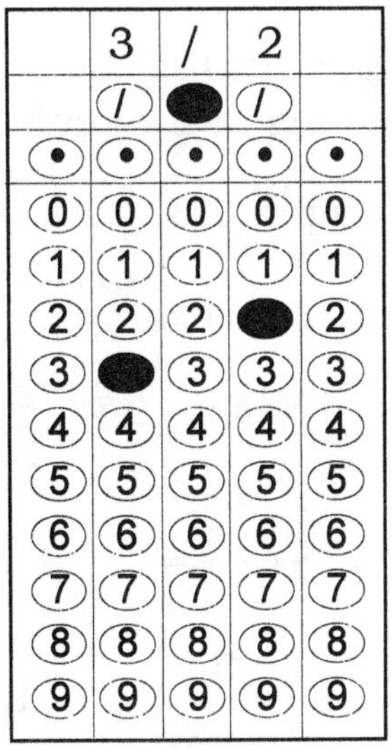

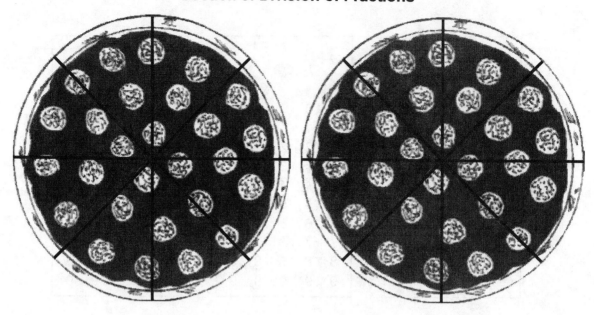

If the rule for dividing fractions is difficult for you to remember, it may be because ever since you were small, when a quantity had to be divided, it became smaller. However, if you divide a quantity by a fraction that is less than one, the result is greater than the original quantity. Think of two pizzas. Each is divided into pieces that are one-eighth the size of the original pie. You start out with two pieces (2 pizzas), and you end up with 16 individual pieces. Put another way, 2 divided by $\frac{1}{8}$ equals 16.

The rule for dividing by a fraction is to invert the fraction you are dividing by and proceed with the problem as if it were a multiplication problem.

In our example, $2 \div \frac{1}{8} = 2 \times 8 = 16$

Let's try another example.

Example: Divide $\frac{4}{7}$ by $\frac{3}{5}$.

Solution: $\frac{4}{7} \div \frac{3}{5} =$ The divisor must be inverted. $\frac{3}{5}$ is changed to $\frac{5}{3}$.

$\frac{4}{7} \cdot \frac{5}{3} =$ Then you proceed with the multiplication problem.

$\frac{(4)(5)}{(7)(3)} = \frac{20}{21}$

Key Point

Try another example.

Example: Solve $\dfrac{4}{11} \div \dfrac{6}{22}$

Solution: $\dfrac{4}{11} \div \dfrac{6}{22} =$ Invert the divisor.

$\dfrac{4}{11} \cdot \dfrac{22}{6} =$ Cancel where possible. 11 divides into 11 and 22.

2 divides into 4 and 6.

$\dfrac{\overset{2}{4}}{\underset{1}{11}} \cdot \dfrac{\overset{2}{22}}{\underset{3}{6}} = \dfrac{2 \times 2}{1 \times 3} = \dfrac{4}{3} = 1\dfrac{1}{3}$

Here is a different type of division problem.

Example: Reduce $\dfrac{\frac{7}{12}}{\frac{21}{8}}$

Solution: Remember, the line that separates the numerator and denominator of a fraction stands for division. Therefore, the problem you are faced with is dividing $\dfrac{7}{12}$ by $\dfrac{21}{8}$.

$\dfrac{7}{12} \div \dfrac{21}{8} =$

$\dfrac{7}{12} \cdot \dfrac{8}{21} =$

$\dfrac{\overset{1}{7}}{\underset{3}{12}} \cdot \dfrac{\overset{2}{8}}{\underset{3}{21}} = \dfrac{1 \times 2}{3 \times 3} = \dfrac{2}{9}$

Division of Fraction Problems

1. $\dfrac{11}{12} \div \dfrac{3}{4}$ 2. $\dfrac{6}{7} \div \dfrac{4}{9}$

3. $\dfrac{2}{13} \div \dfrac{8}{11}$ 4. $\dfrac{15}{16} \div 4\dfrac{1}{2}$

5. The length of the wall in Bob's bedroom is $12\frac{1}{2}$ feet. Along this wall he plans to put a dresser measuring 3 feet 11 inches and a desk to hold his computer and printer. The desk measures 6 feet 5 inches. What is the maximum distance these two pieces of furniture can be apart?

6. Using the information from problem #5, how far apart (in inches) will the two pieces of furniture be if Bob wants an equal distance from the wall to the dresser, between the dresser and the desk, and between the desk and the wall?

Division of Fraction Solutions

1. $\dfrac{11}{12} \div \dfrac{3}{4} = \dfrac{11}{12} \times \dfrac{4}{3} = \dfrac{11}{\underset{3}{\cancel{12}}} \times \dfrac{\overset{1}{\cancel{4}}}{3} = \dfrac{11}{9} = 1\dfrac{2}{9}$

2. $\dfrac{6}{7} \div \dfrac{4}{9} = \dfrac{6}{7} \times \dfrac{9}{4} = \dfrac{\overset{3}{\cancel{6}}}{7} \times \dfrac{9}{\underset{2}{\cancel{4}}} = \dfrac{27}{14} = 1\dfrac{13}{14}$

3. $\dfrac{2}{13} \div \dfrac{8}{11} = \dfrac{2}{13} \times \dfrac{11}{8} = \dfrac{\overset{1}{\cancel{2}}}{13} \times \dfrac{11}{\underset{4}{\cancel{8}}} = \dfrac{11}{52}$

4. $\dfrac{15}{16} \div 4\dfrac{1}{2} = \dfrac{15}{16} \div \dfrac{9}{2} = \dfrac{15}{16} \times \dfrac{2}{9} = \dfrac{\overset{5}{\cancel{15}}}{\underset{8}{\cancel{16}}} \times \dfrac{\overset{1}{\cancel{2}}}{\underset{3}{\cancel{9}}} = \dfrac{5}{24}$

5. The dresser is 3'11" and the computer desk is 6'5". This is a total length of 9'16" or, more correctly stated, 10'4". If we subtract 10'4" from the total length of the wall (12' 6") the difference (2' 2") is the maximum distance that the two pieces of furniture can be apart.

6. In problem 5 we discovered that the total unused space was 2 feet 2 inches. The problem is now to divide the space into 3 equal pieces.

2 feet 2 inches = 26 inches

$26 \div 3 = 26 \times \dfrac{1}{3} = \dfrac{26}{3} = 8\dfrac{2}{3}$ inches

Section 9: Ratios

A ratio is a comparison of two quantities and is an expression of their relative size. You can encounter ratios at a school board meeting: "There are 24 students for every one teacher"; In currency exchange: "You will get 9 German Marks for every 4 U.S. Dollars"; The breakdown of the job force: "There are 2 women for every 3 men".

A ratio can be expressed as a fraction $\frac{2}{3}$, or using a colon – 2:3. Order is important. Obviously, $\frac{2}{3}$ and $\frac{3}{2}$ are not equal. Nor is 2:3 and 3:2.

Example: 24 : 9 is equivalent to which of the following:

    1) 3:2   2) $\frac{15}{3}$ 3) $\frac{8}{3}$ 4) 12:3   5) $\frac{2}{3}$

Solution: Both 24 and 9 can be divided by 3, so the ratio can be reduced to 8:3, or $\frac{8}{3}$. "3)" is the correct answer.

Example: Nancy is 18 years older than her 8 year old daughter. What will the ratio of Nancy's age to her daughter's age be in 6 years?

Solution: Nancy is currently 26 years old. She will be 32 in 6 years. Her daughter will be 14. The ratio will be 32 : 14 or 16 : 7.

Section 10: Proportions

A proportion is a relationship that expresses two equal ratios. From our previous example, we know that:

$$\frac{24}{9} = \frac{8}{3}$$

24 and 9 are in the same ratio as 8 and 3.

In the proportion above, 24 is the first term of the proportion, 9 is the second, 8 is the third, and 3 is the fourth. The second and third terms (9 and 8) are called the means and the first and fourth terms (24 and 3) are called the extremes.

In a proportion, the product of the means is equal to the product of the extremes. An easier way to remember this is the expression "cross-multiplying".

$$\frac{24}{9} \bowtie \frac{8}{3}$$ The product of the means, 9 * 8, is equal to the product of the extremes, 24 * 3. They're both equal to 72.

Example: Find the third term of the proportion $\frac{3}{5} = \frac{?}{25}$.

Solution: 3 * 25 = 5 * ?

75 = 5 * ?

75 = 5 * **15**

$$\frac{3}{5} = \frac{15}{25}$$

Example: Find the fourth term of the proportion $\frac{5}{6} = \frac{15}{?}$

Solution: 5 * ? = 6 * 15

5 * ? = 90

5 * **18** = 90

$$\frac{5}{6} = \frac{15}{18}$$

Ratio and Proportion Problems

1.   Simplify the ratio 28 : 6

2.   The ratio 18 : 12 is equivalent to which of the following?

3. Find the third term of the proportion $\dfrac{3}{8} = \dfrac{?}{24}$.

4. Find the fourth term of the proportion $\dfrac{7}{9} = \dfrac{35}{?}$

5. Find the first term of the proportion $\dfrac{?}{4} = \dfrac{18}{24}$.

6. Find the second term of the proportion $\dfrac{11}{?} = \dfrac{55}{15}$.

7. Edmund, at age 54, is 48 years older than his daughter, Katie. What is the ratio of Edmund's age to his daughter's age? What will the ratio be in 6 years?

8. Gail spends $\dfrac{1}{4}$ of her monthly take home pay on rent, $\dfrac{1}{8}$ on food and clothes, $\dfrac{3}{40}$ on heat and electric, $\dfrac{1}{5}$ on insurance and repayment of loans, $\dfrac{1}{10}$ on transportation, and $\dfrac{1}{16}$ on phone, water, other bills, and miscellaneous household items. What is the ratio of rent expense to heat and electric? If Gail spends \$1,200 on rent, how much does she allow for heat and electric?

9. Daniel is allowed to watch 2 hours of television for every 3 hours he puts in studying. On the average, Daniel studies 10 hours each week. On the average, how much TV time is he allowed?

Ratios and Proportions Solutions

1. The numbers 28 and 6 are divisible by 2.

   28 : 6 = 14 : 3.

2. Eliminate "4" and "5" because the larger number is not first. The numbers 18 and 12 are divisibly by 6.

   18 : 12 = 3 : 2. The answer is "2."

3. $\dfrac{3}{8} \diagtimes \dfrac{?}{24}$ 3 x 24 = 8 x ?   72 = 8 x ?   72 = 8 x 9   ? = 9

4. $\dfrac{7}{9} \diagtimes \dfrac{35}{?}$ 7 x ? = 9 x 35   7 x ? = 315   7 x 45 = 315   ? = 45

5. $\dfrac{?}{4} \diagdown \dfrac{18}{24}$ $? \times 24 = 4 \times 18$ $24 \times ? = 72$ $24 \times 3 = 72$ $? = 3$

6. $\dfrac{11}{?} \diagdown \dfrac{55}{15}$ $11 \times 15 = ? \times 55$ $165 = ? \times 55$ $165 = 3 \times 55$ $? = 3$

7. **Part One:** If Edmund is 48 years older than his daughter, his daughter is 6 years old. The correct ratio is 54:6 or 9:1.

   Part Two: In six years, Edmund will be 60 years old and his daughter will be 12. The correct ratio is 60:12 or 5:1.

8. **Part One:** The two fractions we are comparing are $\dfrac{1}{4}$ and $\dfrac{3}{40}$. One approach would be to write both fractions in terms of the common denominator, 40. The result is $\dfrac{10}{40}$ and $\dfrac{3}{40}$. Out of every 40 dollars spent, 10 dollars is spent on rent and 3 dollars is spent on heat and electric. The ratio is 10:3.

   Another approach is to remember that a ratio is a division. We are dividing $\dfrac{1}{4}$ by $\dfrac{3}{40}$. This is the same as multiplying $\dfrac{1}{4}$ by $\dfrac{40}{3}$.

   $$\dfrac{1}{\overset{1}{4}} \times \dfrac{\overset{10}{40}}{3} = \dfrac{10}{3}$$

   Part Two: The proportion is: 10 is to 3 as 1200 is to what number? We have to solve for the fourth member of the proportion.

   $$\dfrac{10}{3} \diagdown \dfrac{1200}{?}$$

   10 goes into 1200, 120 times

   120 times 3 = 360

   ? = 360

9. The ratio of study time to TV time is 3 : 2.

   3 is to 2 as 10 is to what number?

   $\dfrac{3}{2} = \dfrac{10}{?}$ $3 \times ? = 2 \times 10$ $3 \times ? = 20$ $? = \dfrac{20}{3}$ or $6\dfrac{2}{3}$ hours.

The decimal form of a number is something that you are familiar with because you work with money on a daily basis and dollars and cents use a decimal format. We'll see that decimals are easier to work with than fractions, but it is important that you can work with both and be able to change from one format to the other.

First, let's look at some basic equivalents between decimal form and fractional form of numbers.

Decimal	Definition	Fraction
.1	1 tenth	$\frac{1}{10}$
.01	1 one hundredth	$\frac{1}{100}$
.001	1 one thousandth	$\frac{1}{1000}$

and so on.

Section 1: Changing From Decimal Form to Fraction Form

Changing from decimal form to fraction form is easy. Use the following steps:

Step 1:   The numerator of the fraction will be the decimal number.

Step 2:   The denominator will be the number one followed by a number of zeros equal to the number of decimal places in the original decimal number.

Example: Change .63 to a fraction.

Solution: Step 1:   The numerator will be 63.

Step 2:   There are two decimal places in the fraction .63. The denominator will be the number one with 2 zeros (100).

Therefore, $.63 = \frac{63}{100}$.

If you are changing a decimal to a fraction, and there is a whole number to the left of the decimal, the result will be a mixed number.

Example: Write 7.91 as a fraction.

Solution: 7.91 has a whole number part (7) and a decimal part (.91). Your answer will have a whole number part and a fraction part. The decimal is written as $\frac{91}{100}$ and written to the right of the whole number.

Therefore, $7.91 = 7\frac{91}{100}$

Example: Change .231 to a fraction.

Solution: The numerator will be 231. Since there are three places or digits to the right of the decimal, the denominator of the fraction will be the number one followed by 3 zeros.

Therefore, $.231 = \frac{231}{1000}$

Example: Change 3.926 to a fraction.

Solution: The whole number, 3, is written to the left of the fraction whose numerator is 926 and whose denominator is equal to 1, followed by 3 zeros.

Therefore, $3.926 = 3\frac{926}{1000}$

This fraction can be reduced by dividing both the numerator and denominator by 2.

$$3.926 = 3\frac{926}{1000} = 3\frac{463}{500}$$

Section 2: Changing From Fraction Form to Decimal Form

Key Point

Changing from a fraction to a decimal is a matter of dividing the denominator into the numerator

Example: Change $\frac{3}{4}$ to a decimal.

Solution: $\frac{3}{4} = 3 \div 4 = .75$

Example: Which of the following is equivalent to $\frac{3}{8}$?

1)   .25   2)   .5   3)   .625   4)   .375   5)   .1667

Solution: $\frac{3}{8} = 8\overline{)3.000}$
$$
\begin{array}{r}
.375 \\
8\overline{)3.000} \\
\underline{2\,4} \\
60 \\
\underline{56} \\
40 \\
\underline{40} \\
0
\end{array}
$$

The decimal equivalent of $\frac{3}{8}$ is .375. The correct answer is 4).

Converting Decimals and Fractions Problems

Directions: **Change the following decimals to fractions.**

1. .369   2. 7.25

3. .024   4. 13.8

5. 4.76   6. .005

Directions: **Change the following fractions to decimals.**

7. $\frac{7}{8}$ 8. $\frac{1}{4}$

9. $\frac{4}{5}$ 10. $\frac{9}{2}$

11. $\frac{19}{25}$ 12. $4\frac{7}{10}$

13. $\frac{18}{20}$ 14. $12\frac{3}{8}$

49

Converting Decimals and Fractions Solutions

1. $.369 = \dfrac{369}{1000}$

2. $7.25 = \dfrac{725}{100}$ or $7\dfrac{25}{100} = 7\dfrac{1}{4}$

3. $.024 = \dfrac{24}{1000} = \dfrac{3}{125}$

4. $13.8 = \dfrac{138}{10}$ or $13\dfrac{8}{10} = 13\dfrac{4}{5}$

5. $4.76 = \dfrac{476}{100}$ or $4\dfrac{76}{100} = 4\dfrac{19}{25}$

6. $0.005 = \dfrac{5}{1000} = \dfrac{1}{200}$

7. $\dfrac{7}{8} = 8\overline{)7.000}^{.875} = .875$

8. $\dfrac{1}{4} = 4\overline{)1.00}^{.25} = .25$

9. $\dfrac{4}{5} = 5\overline{)4.0}^{.8} = .8$

10. $\dfrac{9}{2} = 2\overline{)9.0}^{4.5} = 4.5$

11. $\dfrac{19}{25} = 25\overline{)19.00}^{.76} = .76$

A quick way to do this problem is to multiply the numerator and the denominator by 4.

$$\dfrac{19}{25} = \dfrac{19 \times 4}{25 \times 4} = \dfrac{76}{100} = .76$$

If your denominator is equal to 10, 100, 1000 etc., you only need to shift the decimal point in the numerator. Shift the decimal to the left the number of places equal to the number of zeros in the denominator.

12 $\dfrac{7}{10} = .7;\quad 4\dfrac{7}{10} = 4.7$

13. Reduce $\frac{18}{20}$ to $\frac{9}{10}$. Then, just move the decimal point one place to the left.

$$\frac{18}{20} = .9$$

14. $\dfrac{3}{8} = 8\overline{)3.000}^{.375} = .375$; $12\dfrac{3}{8} = 12.375$

Section 3: Rounding Decimals

As we've already discussed, the first position to the right of the decimal is called the tenths position, the second position to the right of the decimal is called the hundredths position, and the third position to the right of the decimal is called the thousandth position. It continues from there, ten thousandth, hundred thousandth, etc.

Key Point

If we want to round a decimal number, say .375 to the nearest hundredth, we are basically asking is .375 closer to .38 or .37. In order to do this we must examine the digit to the right of the hundredth position. This would be the thousandth position. If this digit is 5 or greater, we will add 1 to the hundredth digit and drop the thousandth digit. If this digit is 4 or less, we leave the hundredth position as is and drop the thousandth digit. In our example, .375, the thousandth digit is 5 or greater, so we increase the hundredth digit by 1 to 8, and drop the 5.

.375 rounded to the nearest hundredth is **.38**

.375 rounded to the nearest tenth is **.4**

Example: Round 17.29 to the nearest whole number.

Solution: In this problem you are asked if 17.29 is closer to 17 or 18. You are rounding to the "ones" position, so we look at the digit immediately to the right. That digit is 2. Since it is 4 or less, we leave the ones digit as is and drop all the digits to the right. 17.29 rounded to the nearest whole number is 17.

Section 4: Addition and Subtraction of Decimals

Key Point

The one thing to keep in mind when adding decimal numbers is that you form a column of the decimals that you wish to add. The decimal points must line up one under the other.

Example: Add .19 + 2.6 + .384

Solution:
```
  .190
 2.600
+ .384
-------
 3.174
```

Key Point

Subtraction must be handled in the same way as addition. Line up the decimal points.

Example: Subtract 4.974 from 8.26.

Solution:
```
  8.260
- 4.974
-------
  3.286
```

Example: 17.29 - 3.694, rounded to the nearest hundredth

    1) 13.596   2) 13.71   3) 13.50   4) 13.59   5) 13.60

Solution:
```
 17.290
- 3.694
-------
 13.596
```

The correct answer is 5).

Adding and Subtracting Decimals Problems

Directions: Combine the following decimals.

1. 3.8 + .297 + .55 =

2. 9.21 + 6.592 =

3. 12.16 + 9.5 =

4. 629.86 + 318.6 =

5. 6.2 + 18.4 + .39 =

6. 5.61 – 3.4 =

7. 18.18 – 9.9 =

8. 462.49 – 289.64 =

9. .751 - .026 =

10. 9.078 – 8.64 =

11. **On May 9th, her birthday, Holly deposited $1,446.94 in her checking account. She made another deposit of the same amount at the end of the month. On the 7th of May, she wrote a check for $91.20 to cover her phone bill and another for $72.77 to cover her electric. On the 26th of the month, she wrote checks to cover her charge account (269.20), her monthly bill for car insurance ($123.50), and her next month's rent ($1,125.00). If her beginning balance on May 1st was $319.57, how much money did she have in her account on May 20th? On May 31st? Does Holly have enough money in her account to pay $668.70 for an oil delivery on May 26th?**

Adding and Subtracting Decimal Solutions

1.
```
  3.800
   .297
+ .550
  4.647
```

2.
```
   9.210
 + 6.592
  15.802
```

3.
```
  12.16
+ 9.50
  21.66
```

4.
```
  629.86
+ 318.60
  948.46
```

5.
```
   6.20
  18.40
+  .39
  24.99
```

6.
```
  5.61
- 3.40
  2.21
```

7.
```
  18.18
- 9.90
  8.28
```

8.
```
  462.49
- 289.64
  172.85
```

9.
```
  .751
- .026
  .725
```

10.　　　9.078
　　　 - 8.640
　　　　 .438

11.　Because this is a complicated problem, you may find it helpful to first set up a schedule listing dates bills and deposits are made and include the questions you need to answer. Also, this is an excellent problem to practice using your calculator.

May 1 - $319.57 balance

May 7 - $ 91.20 paid phone

　　　　　$ 72.77 paid electric

May 9 - $1,446.94 deposited in account

May 20 - How much is in the account?

May 26 - $269.20 paid charge account

　　　　　$123.50 paid insurance

　　　　　$1,125.00 paid rent

　　　　　Does Holly have enough in his account to pay the $668.70 oil bill?

May 31 - $1,446.94 deposited in account

　　　　　How much is in the account on this date?

Before Holly made the deposit, she wrote two checks:

　　　　　$91.20　　(Phone)
　　　　+ 72.77　　(Electric)
　　　　　$163.97

That reduced her account balance by $163.97. Her beginning balance was $319.57. Subtracting $163.97 from $319.57, that left her with $155.60 in her account.

　　　　　$319.57
　　　　 - 163.97
　　　　　$155.60

Later that week she deposited $1,446.94, increasing her balance to $1,602.54.

　　　　　$155.60
　　　　 + 1446.94
　　　　　$1602.54　　　(May 20th balance)

After depositing her paycheck, she wrote checks totaling $1,517.70.

$$\begin{array}{ll} \$269.20 & \text{(Charges)} \\ 123.50 & \text{(Insurance)} \\ \underline{+1125.00} & \text{(Rent)} \\ \$1517.70 & \end{array}$$

This reduced her account balance to $84.84.

$$\begin{array}{r} \$1602.54 \\ \underline{-1517.70} \\ \$84.84 \end{array}$$

Therefore, she cannot pay her oil bill until the end of the month when she deposits an additional $1,446.94. Then her balance will be $1,531.78.

$$\begin{array}{r} \$84.84 \\ \underline{+1446.94} \\ \$1531.78 \end{array} \qquad \text{(May 31}^{\text{st}} \text{ balance)}$$

Section 5: Multiplication and Division of Decimals

Multiplication

Key Point

In multiplication, the important point to remember is the number of decimal places in your answer must be equal to the sum of the decimal places in your problem

Example: Find the product of 3.6 and 5.4

Solution: There is 1 decimal place in the first number and 1 in the second. Therefore, the product will have 2 decimal places.

$$\begin{array}{r} 3.6 \\ \times\,5.4 \\ \hline 144 \\ 1800 \\ \hline 19.44 \end{array}$$

Example: The product of 9.1 and 7.9 is:
   1) 112.69
   2) 71.89
   3) 73.169
   4) 42.89
   5) 73.64

Solution:
$$\begin{array}{r} 9.1 \\ \times\,7.9 \\ \hline 819 \\ 6370 \\ \hline 71.89 \end{array}$$

Key Point

Apply relative size of the numbers to the problem. This will give you the relative size of the answer. You are multiplying a number that is a little larger than 9 by a number that is a little less than 8. The answer should be about 72. That would allow you to eliminate answer choices "1" and "4". Choice "3" can also be eliminated because there will be only 2 decimal places in the answer, and choice "5" can be eliminated because the answer must end with the number 9 (9 x 1 = 9). Therefore choice "2" can be shown to be the correct answer without even multiplying the problem out. Being able to notice relative size and looking at the decimal answers can save you a large amount of time on the test.

Division

Key Point

> When dividing by a decimal, shift the decimal point as many places to the right as is necessary to make the divisor (the number you are dividing by) a whole number. You must then shift the decimal point of the dividend (the number you are dividing into) an equal number of places. The decimal point in the quotient (the answer) is placed immediately above the decimal in the dividend.

Example: Divide .00105 by 3.5

Solution: $3.5\overline{).00105}$

$$
\begin{array}{r}
.0003 \\
35\overline{).0105} \\
\underline{105} \\
0
\end{array}
$$

Example: Divide 25 by 3.21 and round your answer to one decimal place

Solution: $3.21\overline{)25}$

$$
\begin{array}{r}
7.78 \\
321\overline{)2500.00} \\
\underline{2247} \\
2530 \\
\underline{2247} \\
2830 \\
2568
\end{array}
$$

To round your answer to the nearest tenth, examine the digit to it's right which is 8. It is 5 or greater, so increase 7 by 1 and drop the remaining decimals. The rounded answer is 7.8

Multiplying and Dividing Decimal Problems

Directions: **Multiply the following decimals.**

1. (3.29) (1.8)

2. 16.2 x .23

3. (12.29) (7.42)

4. .091 x 3.28

5. (110.4) (8.1)

Directions: **Divide the following decimals**

6. .0105 by 3.5

7. 17.08 by 2.8

8. 62.985 by 16.15

9. 77.361 by 21.4

10. 117 by 8.3 rounded to 2 decimal places

Multiplying and Dividing Decimal Solutions

1.
```
    3.29
  x 1.8
   2632
   3290
  5.922
```

2.
```
   16.2
  x .23
    486
   3240
  3.726
```

3.
```
   12.29
  x 7.42
    2458
   49160
  860300
  91.1918
```

4.
```
    3.28
  x .091
     328
    2952
  .29848
```

5.
```
   110.4
  x 8.1
    1104
   88320
  894.24
```

6.
```
                    .003
  3.5).0105 = 35).105
                    105
                      0
```

7.
```
                    6.1
  2.8)17.08 = 28)170.8
                  186
                   28
                   28
                    0
```

8.
```
                        3.9
  16.15)62.985 = 16.15)6298.5
                       4545
                      14535
                      14535
                          0
```

9.
```
                      3.615
  21.4)77.361 = 214)773.610
                    642
                    1316
                    128 4
                      3 21
                      2 14
                      1070
                      1070
                         0
```

60

10. $8.3\overline{)117}$ = 83$\overline{)1170.000}$ = 14.10 rounded to 2 decimal places.

$$
\begin{array}{r}
14.096 \\
83\overline{)1170.000} \\
\underline{83} \\
340 \\
\underline{332} \\
800 \\
747 \\
530 \\
498 \\
\end{array}
$$

Section 1: Changing Between Decimal and Percent Form

Changing from a Decimals to a Percent

Key Point

Percent is another way that you can refer to a part of 100. 1 percent is 1 part of 100. 100 percent is 100 parts of 100, or the entire quantity. The symbol for percent is "%"

Changing from decimal form to percent form is just a matter of shifting the decimal point two places to the right and adding the percent symbol.

For example:

.25 = 25%   .325 = 32.5%

Changing from a Percent to a Decimal

Key Point

In problems where you want to find a percent of a number, you will perform a multiplication. But in order to do that, you will have to convert the percent into a decimal. You can probably see that changing from a percent to a decimal is just as easy as changing from a decimal to a percent – just reverse the procedure. In other words, shift the decimal point two places to the left and drop the percent symbol.

For example:

62% = .62   35.4% = .354   45½% = 45.5% = .455

Changing Decimals and Percents Problems

Change the following decimals to percents.

1. .67   2. .0031   3. 5.97   4. .801

5. .093

Change the following percents to decimals.

6. 50%   7. 32 1/4%   8. 6.4%   9. .098%

10. 320%

Changing Decimals and Percents Solutions

1. .67 = .67.% = 67%

2. .0031 = .00.31 = .31%

3. 5.97 = 5.97.% = 597%

4. .801 = .80.1% = 80.1%

5. .093 = .09.3% = 9.3%

6. 50% = .50. = .5

7. 32 1/4% = 32.25% = .32.25 = .3225

8. 6.4% = .06.4 = .064

9. .098% = .00.098 = .00098

10. 320% = 3.20. = 3.2

Section 2: Changing Between Fraction and Percent Form

Changing from a Fraction to a Percent

Key Point

In order to change a fraction to a percent, you must first change the fraction to decimal form. Once you've done that, you have, as mathematicians like to point out, reduced it to a previously solved problem. That being, changing a decimal to a percent.

Example: Change $\dfrac{5}{8}$ to a percent

Solution: $\dfrac{5}{8} = 8\overline{)5.000}^{.625} = 62.5\%$

Changing from a Percent to a Fraction

Key Point

The % symbol expresses a division by 100. You can drop it as long as you express that division some other way. One way to show that 100 divides a number is to write the number as the numerator of a fraction with 100 as the denominator.

Example: Write 72% as a fraction.

Solution: $72\% = \dfrac{72}{100} = \dfrac{18}{25}$

Example: Write 12.5% as a fraction.

Solution: $12.5\% = \dfrac{12.5}{100} = \dfrac{125}{1000} = \dfrac{1}{8}$

Example: Change $4\dfrac{1}{6}\%$ to a fraction.

Solution: $4\dfrac{1}{6} = \dfrac{25}{6}\% = \dfrac{25}{600} = \dfrac{1}{24}$

Changing Fraction and Percent Problems

Change the following fractions to percents.

1. $\dfrac{2}{5}$ 2. $\dfrac{3}{8}$ 3. $3\dfrac{1}{4}$ 4. $\dfrac{17}{20}$

5. $\dfrac{9}{25}$

Change the following percents to fractions.

6. 27%   7. 9.4%   8. 37 1/2%   9. 7 1/4%

10. 120%

Changing Fraction and Percent Solutions

1. $\dfrac{2}{5} = 5\overline{)2.0}^{.4}$

   .4 = .40.% = 40%

2. $\dfrac{3}{8} = 8\overline{)3.000}^{.375}$

   .375 = .37.5% = 37.5%

3. $3\dfrac{1}{4}$ = 3.25 = 3.25.% = 325%

4. $\dfrac{17}{20} = 20\overline{)17.00}^{.85}$

   .85 = .85.% = 85%

5. $\dfrac{9}{25} = \dfrac{9\times4}{25\times4} = \dfrac{36}{100}$ = .36 = .36.% = 36%

6. 27% = $\dfrac{27}{1\times100} = \dfrac{27}{100}$

7. 9.4% = $\dfrac{94}{10\times100} = \dfrac{94}{1000} = \dfrac{47}{500}$

8. $37\dfrac{1}{2}$% = $\dfrac{75}{2\times100} = \dfrac{75}{200} = \dfrac{3}{8}$

9. $7\dfrac{1}{4}$% = $\dfrac{29}{4\times100} = \dfrac{29}{400}$

10. 120% = $\dfrac{120}{1\times100} = \dfrac{120}{100} = \dfrac{6}{5} = 1\dfrac{1}{5}$

Section 3: Word Problems Dealing With Percent

Taking A Percent of a Number

There are basically three types of percent word problems. The first type we will discuss is taking the percent of a number. You may find it the easiest of the three types because it is one you work with the most often in everyday life.

Key Point

> The word **of** in English is translated into the arithmetic operation multiplication. The phrase "20% of 165" is translated into 20% times 165.

In order to perform the multiplication, you must first rewrite 20% as a decimal.

20% = .20

```
  165        First change the percent to the decimal equivalent form.
x .20
33.00        20% = .2    20% of 165 = 33
```

Example: Find 15% of 70

Solution: 15% = .15

```
     70
   x .15
    350
    700
   10.5
```

15% of 70 = 10.5

Now, look at a very common type of situation in which you will have to use this skill – when you need to calculate sales tax on an expensive item.

Example: Bill is thinking of buying a riding mower that costs $2,250. Bill lives in New Jersey where the sales tax is 6%. What is the total cost of the mower?

Solution: The sales tax is 6% of $2,250. In arithmetic, the word "of" translates to multiplication.

$2250 * 6% = $2250 * .06 = $135.00 The sales tax is $135.00.

The total cost is $2,250 + $135.00 = $2,385.

Example: Calculate a 20% tip on a $55.00 dinner.

Solution: Again, our answer is the product of the original number, 55, and the percentage, 20%. The easiest way to do this is first to find 10% of the number and double that answer, since 20% is twice 10%. We use 10% because it is easy to calculate. Multiplying by 10% simply shifts the decimal point one place to the left of the number.

$$55 \times 10\% = 5.5$$

When we double this result, we get 11. 20% of $55.00 is $11.00.

Section 4: Expressing a Number as a Percent of Another

Key Point

The second type of percent word problem is one in which you need to express one number as a percent of another. For example, finding what percentage 24 is of 96. As a fraction, this would be written as $\frac{24}{96}$. The **part** is always placed in the **numerator**, and the **whole** is always in the **denominator**. To get a decimal, you would divide 96 into 24 and get .25. This result is more easily obtained if you first reduce $\frac{24}{96}$ to $\frac{1}{4}$.

Shifting the decimal point two places to the right gives the desired result of 25%.

Example: If you earn \$165 on a \$3000 savings account what percentage interest does this represent?

Solution: Restating the question yields, "What percent of 3000 is 165?"

$$\frac{\text{Part} \rightarrow}{\text{Whole} \rightarrow} \frac{165}{3000} = \frac{165 \div 5}{3000 \div 5} = \frac{33}{600} = \frac{11}{200}$$

$$200 \overline{)11.000} = 5.5\%$$
$$\begin{array}{r} .055 \\ \underline{10\,00} \\ 1\,000 \\ \underline{1\,000} \\ 0 \end{array}$$

Example: During the 2000 – 2001 season, the nationally ranked Rutgers University women's basketball team won 23 games and lost 8. What was their winning percentage rounded to the nearest tenth of a percent?

Solution: The number of games won was 23. The total number of games played was 23 plus 8, or 31. Using your calculator to help with the division, the percentage won was:

23/31 = .7419 = 74.19%

Rounded to the nearest tenth = 74.2%

Section 5: One Number Represents a Percentage of an Unknown Number

Key Point

> The third type of percentage word problem is perhaps the most confusing for people. It is the type in which you are told that one number represents a percentage of an unknown number and you are asked to find the original number.

Example: 15 is 20% of what number.

Solution: First translate the English into math.

15 = 20% times some number.

To find the unknown number we must reverse the process. You must divide the answer 15 by 20%.

$15 \div 20\% = 15 \div .20$

$$
\begin{array}{r}
75. \\
.2\overline{)15.0} \\
14 \\
\overline{10} \\
10 \\
\overline{0}
\end{array}
$$

15 is 20% of 75

Example: A store is running a 35% sale on all jewelry. A pair of earrings is on sale for $55.25. What was the original price of the earrings? How much money would you save buying the item on sale?

Solution: When you know the percentage and know the "answer", you find the original number by dividing the answer by the percentage. Since there was a 35% sale, we know that $55.25 represents 65% of the original number. Use your calculator to help you with the following division:

$55.25 \div .65 = 85.$

The original price was $85.00. You would save the difference between $85.00 and $55.25, or $29.75.

69

Percentage Word Problems

1. Find 17% of 64.

2. Find 160% of 49.

3. Find 5% of $820.

4. Find the interest on a $480 loan if the interest charge is 10%.

5. Calculate a 20% tip on a $15.00 dinner.

6. What percent of 65 is 13?

7. What percent of 90 is 16.2?

8. If a $25.00 late penalty is charged on a $500.00 washer/dryer, what percent is the penalty?

9. A high school basketball team wins 15 games and loses 5. What is their winning percentage?

10. Six is 75% of what number?

11. Twenty-six is 40% of what number?

12. A football team that lost 4 times, lost 25% of their games. How many games did they play?

13. Bob had a 60% success rate with his indoor plants. If 24 survived, how many plants did he start with?

14. A local ski shop announces a 50% sale on all ski jackets in stock. The ad goes on to say that selected items will be marked down an additional 50% at the cash register. If you were to take advantage of the double reduction, what would the final marked down price by on an item that originally sold for $180.00? What was the total dollar discount? What was the total percent discount? What would the total cost to you be if you had to pay 5% sales tax?

15. Mary Lou opens up an IRA account so that she can save money for her retirement. Each year she puts away $2 \frac{3}{4}$% of her annual income and each year her investment grows at $7 \frac{1}{2}$% interest on amounts that were on deposit for a-full year. She earns $20,000 the first year and gets a 10% raise the next year. If she always make her IRA contribution at the end of the year, how much money will be in her account at the end of the second year?

Percent Word Problem Answers

1.
$$
\begin{array}{r}
64 \\
\times\ .17 \\
\hline
448 \\
650 \\
\hline
10.88
\end{array}
$$

2.
$$
\begin{array}{r}
49 \\
\times\ 1.60 \\
\hline
2940 \\
4900 \\
\hline
78.40
\end{array}
$$

3.
$$
\begin{array}{r}
820 \\
\times\ .05 \\
\hline
41.00
\end{array}
$$
   or 10% of 820 = 82.0 $5\% = \dfrac{1}{2} 10\% = 82 \div 2 = 41$

4.
$$
\begin{array}{r}
480 \\
\times\ .10 \\
\hline
48.00
\end{array}
$$
   or 10% of 480 = 48.0

5.
$$
\begin{array}{r}
15.00 \\
\times\ .20 \\
\hline
3.0000
\end{array}
$$
   or 10% of 15 = 1.5   20% = (2)10% = (1.5)(2) = 3

6. $\dfrac{13}{65} = 65\overline{)13.00}$
$$
\begin{array}{r}
.20 \\
65\overline{)13.00} \\
\underline{13\,0} \\
00
\end{array}
$$ = 20%

7. $\dfrac{16.2}{90} = 90\overline{)16.20}$
$$
\begin{array}{r}
.18 \\
90\overline{)16.20} \\
\underline{9\,0} \\
7\,20 \\
\underline{7\,20} \\
0
\end{array}
$$ = 18%

8. $\dfrac{25}{500} = 500\overline{)25.00}$
$$
\begin{array}{r}
.05 \\
500\overline{)25.00} \\
\underline{25\,00} \\
00
\end{array}
$$ = 5%

9. $\dfrac{15}{15+5} = \dfrac{15}{20} = 20\overline{)15.00}$
$$
\begin{array}{r}
.75 \\
20\overline{)15.00} \\
\underline{14\,0} \\
1\,00 \\
\underline{1\,00} \\
0
\end{array}
$$ = 75%

10. $\dfrac{6}{75\%} = \dfrac{6}{.75} = .75\overline{)6.00} = 75.\overline{)600}$ 6 is 75% of 8

$$\begin{array}{r} 8 \\ 75.\overline{)600} \\ \underline{600} \\ 0 \end{array}$$

11. $\dfrac{26}{40\%} = \dfrac{26}{.4} = .4\overline{)26.0} = 4.\overline{)260}$ 26 is 40% of 65

$$\begin{array}{r} 65 \\ 4.\overline{)260} \\ \underline{24} \\ 20 \\ \underline{20} \\ 0 \end{array}$$

12. $\dfrac{4}{25\%} = \dfrac{4}{.25} = .25\overline{)4.00} = 25.\overline{)400}$ They played 16 games.

$$\begin{array}{r} 16 \\ 25.\overline{)400} \\ \underline{25} \\ 150 \\ \underline{150} \\ 0 \end{array}$$

13. $\dfrac{24}{60\%} = \dfrac{24}{.6} = .6\overline{)24.0} = 6.\overline{)240}$ He started with 40 plants.

$$\begin{array}{r} 40 \\ 6.\overline{)240} \\ \underline{24} \\ 00 \end{array}$$

14. An item selling for $180.00 would initially be marked down to $90.00 (half or 50% of $180.00). At the register, the $90.00 would be reduced again by 50% of $45.00. The final sales price of $45.00 represents a total savings of $135.00 ($180.00 - $45.00).

The total discount was 75% $\dfrac{135}{180} = \dfrac{3}{4}$

10% of $45.00 is $4.50. 5% would be $2.25

$45.00 + $2.25 = $47.25

The total cost of the jacket with tax equals $47.25.

15. The problem as a whole may seem very involved and complicated, but if we break it into its parts, we can see that it simply involves a series of decimal multiplications and additions.

At the end of the first year, Mary Lou deposits $2\dfrac{3}{4}\%$ of $20,000.

$2\dfrac{3}{4}\% = 2.75\% = .0275$

$$\begin{array}{r} .0275 \\ \times\ 20000 \\ \hline 550.0000 \end{array}$$

With some practice, you could do this in your head. To multiply .0275 by 10,000 you would simply move the decimal 4 places to the right. The result is 275. Multiplying by 20,000 would produce a value twice as large, or 550.

So at the end of the first year the account stands at $550.00

Mary Lou's raise in the second year equals 10% of $20,000 = $2,000. Her salary, therefore, is $22,000. Her contribution at the end of the year equals $605.00.

```
      .0275
    x 22000
     550000
    5500000
    605.0000
```

Key Point

Now do not forget that her first contribution earns interest for a full year.

Her total balance at the end of the second year is:

$550 + 7 $\frac{1}{2}$% interest on $550 + $605

The interest is:

```
     .075
    x 550
     3750
    37500
    41.250
```

The total account is:

$550.00	amount deposited 1st year
41.25	interest earned on 1st year's amount
+ 605.00	amount deposited 2nd year
$1196.25	

Section 6: Using Formulas

For the exam, you will not be required to memorize any formulas. They will be provided for you. However, you will need to understand when it is appropriate to apply a specific formula, and to know how the data in a particular problem relates to the components of the formula.

The following three examples demonstrate this requirement. You can use your calculator on these problems.

Example: It's time for Bob to buy a minivan, and he has already qualified for a $15,000 loan. He finds a car he likes, but the sticker price is $23,750. However, the car company is offering a $1,500 cash rebate, and the dealer is offering a $750 customer loyalty discount. Since Bob is a loyal customer, the dealer also sent him a vehicle allowance voucher in the mail for $955 if he buys a car during their Leadership Celebration next weekend. If Bob wants to put aside $1,000 for sales tax, how much of a trade-in must the dealer offer him in order to make a deal?

    (1) $5,545
    (2) $6,545
    (3) $7,295
    (4) $7,500
    (5) $8,250

Solution: **The correct answer is (2).**

Bob qualified for a $15,000 loan, but wants keep $1000 in reserve, so he only has $14,000 to spend. Subtracting his discounts and rebates from the sticker price of the car, brings the cost down to $20,545.

    $23,750.00
      $1,500.00 (rebate)
        $750.00 (loyalty discount)
 -   $955.00 (voucher)
    $20,545.00

If Bob only has $14,000 to spend, the trade-in has to cover the difference between these two numbers.

    $20,545.00
 - $14,000.00
    $6,545.00 = trade-in value.

Example: Colleen loans $10,000 for 4 years to Michelle. Michelle agrees to pay back the loan and simple interest at an annual rate of $8\frac{3}{4}\%$. How much, in total, will she pay back at the end of the four years?

  (1) $875
  (2) $3,500
  (3) $10,875
  (4) $13,500
  (5) $15,000

Solution: **The correct answer is (4).**

Interest = Principal × Rate × Time

Interest = 10,000 × .0875 × 4

Interest = 3,500

Total payback = Principal + Interest

Total Payback = $10,000 + $3,500 = $13,500

Example: Marissa averages 50 mph on her way to work and makes the trip in 45 minutes. How many miles does she travel to work?

  (1) 12.5
  (2) 25
  (3) 37.5
  (4) 50
  (5) $66\frac{2}{3}$

Solution: **The correct answer is (3).**

Distance = Rate × Time

Distance = $50 \times \frac{3}{4}$ or $50 \times .75$

Distance = 37.5

This is an easy solution to estimate. Half of 50 is 25 and 100% of 50 is 50.

75%, or $\frac{3}{4}$, is half way between these two numbers. You know that the answer is greater than 25, so (1) and (2) must be incorrect. You also know that the answer is less than 50, so (4) and (5) must also be incorrect. The only possible correct answer is (3).

Section One: Word Problems With Decimals and Fractions

The last few problems from the previous chapter serve as a good lead-in to our current discussion. Many of the problems that you will be expected to solve on the exam will not require a formula, but you will need to know what arithmetic operation to perform (addition, subtraction, multiplication, or division) and be able to work with the fraction, decimal, and percent forms that we've already covered. Section 6 of Chapter 3 covered percent word problems. On the following pages, we'll cover some examples of decimals and fractions.

 $1.09

 $.69

Example: A quart of soda normally sells for $.69. A half gallon sells for $1.09. If there are four quarts to a gallon, what is the savings if you buy a gallon of seltzer by the half gallon container instead of individual quarts?

Solution: Two half gallons would cost 2 times $1.09 or $2.18.

If you were to buy individual quarts, you would have to buy 4 to make up a gallon. 4 times $.69 = $2.76.

The savings is $2.76 - $2.18 = $.58.

 $1.09

 Sale $.53

Example: Using the information from the previous problem, if the quarts are on sale this week for 53¢, what is the best buy for a gallon of seltzer?

Solution: Four quarts at 53¢ each would cost a total of $2.12.

You have already determined that two half gallons would cost $2.18.

So, this week, the quarts are the better buy. The savings would be 6 cents ($2.18 – $2.12). Sometimes the savings are not always in the larger container.

Example: It's time for Bob to buy a minivan, and he has already qualified for a $15,000 loan. He finds a car he likes, but the sticker price is $23,750. He has saved $5,000 and the sales tax will be 6%. If he can save $350 per week how long until he can buy the car.
(1) 10 weeks
(2) 11 weeks
(3) 14 weeks
(4) 15 weeks
(5) over a year

Solution: **The correct answer is (4).**

Bob qualified for a $15,000 loan and he currently has $5,000 in savings, so this means he currently had $20,000 to pay for the car. The total cost of the car is equal to the price of the car plus the tax.

$23,750 + (.06)(\$23,750) = \$25,175$

The amount Bob still needs to save is equal to the total price minus the amount he has currently saved.

$25,175 - \$20,000 = \$5,175$

The number of weeks is equal to the amount needed to be saved divided by the amount that is saved each week. This number must then be rounded up to the nearest whole week.

$5,175 \div \$350 = 14.78...$ It will take him 15 weeks.

3.7 pounds   53 oz. $3 \frac{3}{4}$ pounds

Example: If there are 16 ounces to a pound, which is heaviest, a bag weighing 3.7 pounds, one weighing 53 ounces, or one weighing $3 \frac{3}{4}$ pounds?

Solution: Dividing 16 into 53, we convert 53 ounces into $3 \frac{5}{16}$ pounds.

Next, convert the fraction into decimal form:

53 ounces = $3 \frac{5}{16}$ pounds = 3.3125 pounds

$3 \frac{3}{4}$ pounds, written in decimal form, is 3.75

Our third weight, 3.7, is already in decimal form. Obviously, 3.75 is the largest decimal, so $3 \frac{3}{4}$ pounds is the heaviest weight.

Section Two: Arithmetic Formulas

You will not have to memorize formulas for the exam. You'll be provided with a list. You can find a copy of the list at the beginning of this book.

For this section, we'll focus on the last three formulas on the list.

Example: Colleen loans $10,000 for 4 years to Michelle. Michelle agrees to pay back the loan and pay simple interest at an annual rate of 8¾%. How much interest, will she pay back at the end of the four years?
    (1) $875
    (2) $3,500
    (3) $10,875
    (4) $13,500
    (5) $15,000

Solution: **The correct answer is (2).**

Interest = Principal x Rate x Time

Interest = 10,000 x .0875 x 4

Interest = 3,500

Example: Marissa averages 50 mph on her way to work and makes the trip in 45 minutes. How many miles does she travel to work and back?
    (1) 25
    (2) 37.5
    (3) 50
    (4) $66 \frac{2}{3}$
    (5) 75

Solution: **The correct answer is (5).**

Distance = Rate × Time

Distance = $50 \times \frac{3}{4} \times 2$ or 50 x .75 x 2

Distance = 75

Example: If Orlando and Sue can buy a package of 3 infant undershirts for $11.95, how much will it cost to by 15 undershirts?

(1) $59.75
(2) $179.25
(3) $35.85
(4) $11.95
(5) $60.00

Solution: **The correct answer is (1).**

Total Cost = (Number of Units) × (Price Per Unit)

There are 3 undershirts to a package, so if they need 15 undershirts, they will have to buy 5 packages (or Units).

Total Cost = 5 x $11.95 = $59.75

Section Three: The Concept of "Average"

Key Point

Most often when we use the term "average" we are referring to the **Arithmetic Mean**.

To find the **Mean** of a group of numbers, you simply sum all the number values and divide by the number of items you're adding together.

Example: A 10 question quiz resulted in the following scores:

3, 4, 5, 5, 6, 7, 7, 7, 10

What is the mean test score?

Solution: There are 9 test scores, so the number of items is 9. The sum of those 9 items is 54. The mean is equal to 54 ÷ 9 = 6.

Key Point

The **Median** defines a middle point. It is the number value of one of the items we are examining. It is uniquely positioned such that there are an equal number of items whose value is less than the median as those whose value is greater than the median.

Example: Using the example above of our test scores:

3, 4, 5, 5, 6, 7, 7, 7, 10

Solution: 6 would also be the median score. There are 4 quiz scores below 6 and 4 above.

Key Point

The **Mode** of a series of numbers is number value that occurs most frequently within the series.

Example: Using our example one more time, we see that the most common quiz score is 7. It occurs 3 times. The next most frequent value is 5. The remaining test values occur one time each. So, the most frequent value, 7, is the mode of the test scores.

Example: What is the median of the following 9 numbers:

2, 8, 10,10,14,15,18,19, 21

1) 10
2) 13
3) 14
4) 18
5) 21

81

Solution: **The correct answer is 3).**

There are 4 values less than number 14 and four numbers greater than 14, so 14 is the median.

13 is the mean. It's equal to the sum of all the numbers (117) divided by the number of items, which is 9.

10 is the mode. It is the only value that occurs twice.

Example: The difference between the mean and the mode of the numbers:

6, 10, 11, 16, 11, 19, and 11 is:

1) 0
2) 1
3) 11
4) 12
5) 16

Solution: **The correct answer is (2).**

The mean of these numbers is 12, which is equal to the sum (84) divided by the number of items (7).

The mode is the most common value which is also 11, it occurs three times. The difference between these two numbers is 1.

Zero is the difference between the median and the mode. The median is also 11. There are two items that have value less than 11 (6 and 10) and two items that have value greater than 11 (16 and 19).

Chapter 5: Geometric Shapes and Relationships

Our study of Geometry will be covered over two chapters. This chapter will concentrate on a discussion of basic shapes, relationships of sides and angles, and the formulas for perimeter and area. Chapter seven will cover coordinate geometry.

Section 1: Information About Geometric Figures

Before you begin your review of geometric formulas, it is important for you to become familiar with the various types of angles and their relationships to each other. Angles are classified according to the number of degrees they contain. Also, there are certain truths that are constant for different types of geometric figures. Some questions will deal with this type of information and if you know the most common facts, you will find this type of problem easy to solve.

Key Point

The total number of degrees in a circle is 360°.

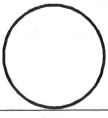

Key Point

Angles are measured in degrees and can be classified according to the number of degrees in them.

An angle whose measure is between 0° and 90° is an **ACUTE** angle.

ACUTE

An angle whose measure is 90° is called a **RIGHT** angle. The symbol to indicate a right angle is ⌐,

RIGHT

Angle whose measure is between 90° and 180° is an **OBTUSE** angle.

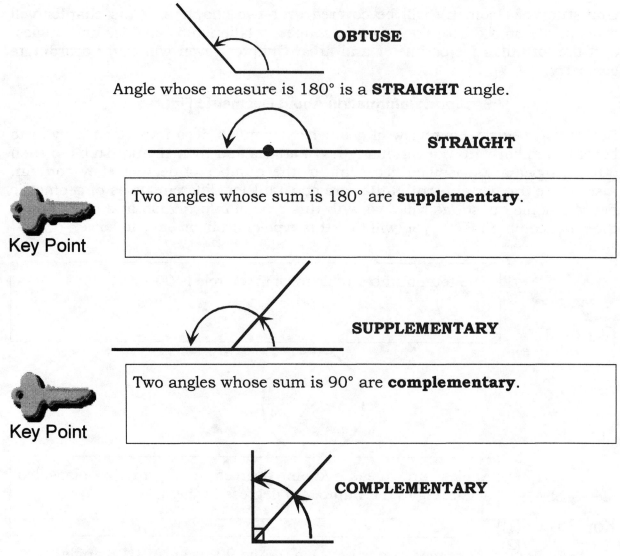

OBTUSE

Angle whose measure is 180° is a **STRAIGHT** angle.

STRAIGHT

Key Point

Two angles whose sum is 180° are **supplementary**.

SUPPLEMENTARY

Key Point

Two angles whose sum is 90° are **complementary**.

COMPLEMENTARY

Let's try some questions. You will be able to answer these questions from the information you have just learned.

Item 1 is based on the following diagram.

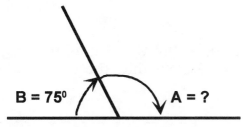

1. **If angle A s 30°, what is the size of angle B?**

        (1)   40°
        (2)   75°
        (3)   60°
        (4)   150°
        (5)   45°

Solution: Because you can see from the diagram that the 2 angles form a right angle and you know that the number of degrees in a right angle is 90°, you can easily pick the correct answer, 60° or number (3). Also, the two angles are complementary, equal 90°. 30 + 60 = 90

Item 2 is based on the following diagram.

2. **If angle B is 75°, what is the size of angle A?**

        (1)   50°
        (2)   130°
        (3)   75°
        (4)   105°
        (5)   85°

Solution: Because you know from the diagram that the 2 angles form a straight angle or are supplementary angles, you would easily pick answer number (4). There are 180° in a straight angle,

75 + 105 = 180.

Let's continue our review.

Key Point

The total number of degrees in the angles of a triangle is 180°.

How might this help you answer a question? Let's take a look.

3.   **In a given triangle, ∠A = 90°, ∠B = 45°. What is the measure of ∠C?**

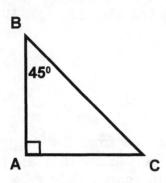

(1)   90°
(2)   45°
(3)   180°
(4)   75°
(50) 60°

Solution: Because you know that the sum of the angles in a triangle is 180°, you know immediately that the correct answer is 45° or number (2).

Key Point

The total number of degrees in a four-sided figure (square, parallelogram, trapezoid rectangle) is 360°

Item 6 is based on the following diagram.

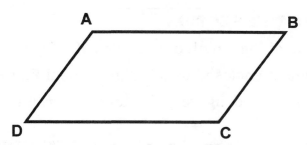

4. **If ∠A = 120°, ∠B = 60°, and ∠C = 120°, how many degrees are in ∠D?**

    (1) 60°
    (2) 45°
    (3) 120°
    (4) 75°
    (5) 80°

Solution: The correct answer is number (1). The total number of degrees in the angles of a parallelogram is 360°. 120 + 60 + 120 = 300.

360 - 300 = 60.

Although there are other facts relating to the relationships of angles and degrees, these are the most commonly tested on the GED Mathematics test. It is important for you to remember these facts because these can be quickly and easily answered.

Section 2: The Rectangle

WORDS TO KNOW IN THIS SECTION

PERIMETER: the sum of the length of the sides of a figure

AREA: size measurement enclosed by a 2 dimensional figure

The rectangle is a four-sided figure, all angles are right angles and opposite sides are equal.

The first formula we will use is one for finding the **PERIMETER** of a rectangle.

Key Point

Perimeter of a rectangle is equal to twice the length, plus twice the width.

P = 2L + 2W

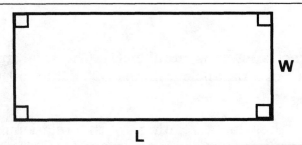

Key Point

Length and width could also be referred to as base and height.

P = 2b + 2h

Example: Find the perimeter of a rectangle whose length is 10 inches and width is 4 inches.

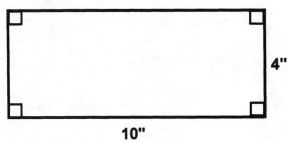

Solution: P = 2l + 2w

P = 2(10") + 2(4")

P = 20" + 8"

P = 28 inches

88

Example: Find the perimeter of a rectangle whose length is 3 inches and width is 8 inches.

Solution: P = 2l + 2w

P = 2(3") + 2(8")

P = 6" + 16"

P = 22 inches

Example: Find the perimeter of a rectangle whose length is 9" and whose width is 3".

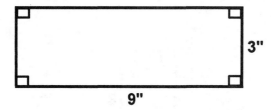

1) 12"
2) 27"
3) 6"
4) 2'
5) 93"

Solution: P = 2(1) + 2(w)

P = 2(9") + 2(3")

P = 18" + 6"

P = 24"

There are 12" in a foot, so 24 inches equals 2 feet. The correct answer is (4), 2'.

Key Point

The second formula we need to discuss is one for finding the **AREA** of a rectangle. The area of a rectangle is equal to length times width. It is written as:

A = l x w

89

This could also be written as A = b x w. b represents base and w represents width.

Key Point

> The unit of measure for area is **square** units.

Example: Find the area of rectangles used in the previous examples.

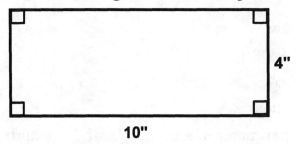

Solution: A = l x w
A = 10" x 4"
A = 40 sq. in.

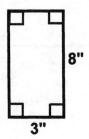

Solution: A = l x w
A = 3" x 8"
A = 24 sq. in.

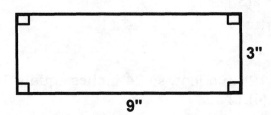

Solution: A = l x w
A = 9" x 3"
A = 27 sq. in.

90

Section 3: The Square

The square is a rectangle with four equal sides.

Key Point

The **PERIMETER** of a square is equal to 4 times the side.
P = 4s

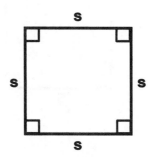

Let us look at the following examples.

Example: Find the perimeter of a square whose side is 6'

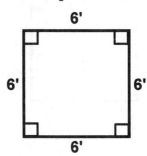

Solution: P = 4s
             P = 4(6')
             P = 24 ft.

Example: Find the perimeter of a square whose side is 1 foot.

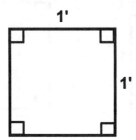

Solution: P = 4s
             P = 4(1")
             P = 4 ft.

Key Point

The **AREA** of a square is equal to side times side, or side squared. The formula is written as:

A = s x s or A = s²

Applying the formula for the area of a square on the squares in previous examples, we get the following:

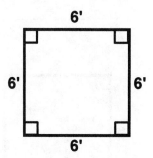

A = s x s
A = 6' x 6'
A = 36 sq. ft.

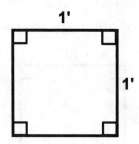

A = s x s
A = 1' x 1'
A = 1 sq. ft.

Here is another example.

Example: Find the number of square inches in a square foot.

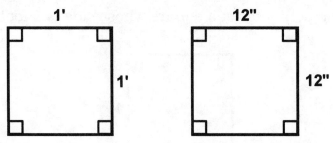

Solution: A = s x s
A = (12")(12")
A = 144 sq. in.

There are 144 square inches in one square foot.

Section 4: The Triangle

A triangle is a three-sided figure.

To begin the review, let's look at the different types of triangles. Knowing the various names and the properties of each will help you answer questions on the GED.

Triangles can be placed into categories according to the type of angles they have.

ACUTE - has three acute angles

RIGHT - has one right angle

OBTUSE - has one obtuse angle

EQUIANGULAR - has three angles of equal measure

Triangles may also be categorized by length of their sides

SCALENE - has no sides the same length

ISOCELES - has 2-sides the same length

EQUILATERAL - all sides are the same length

How can this information help you answer a question on the GED? Let's look.

The diagram refers to question 1.

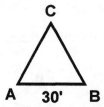

1. **If Triangle ABC is an equilateral triangle and side AB is 30 feet, what is the length of side AC?**

    (1) 45 ft.
    (2) 60 ft.
    (3) 30 ft.
    (4) 50 ft.
    (5) 35 ft.

Solution: Because you know that all sides in an equilateral triangle are equal, you can quickly see the correct answer is, number (3), 30 ft.

Let's continue our review of the formulas used to find area and perimeter.

Key Point

> The **PERIMETER** of a triangle is equal to the sum of the sides.
> P = a + b + c

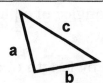

Example: Find the perimeter of a triangle whose sides are 3', 4' and 5'.

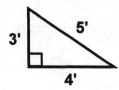

Solution: P = a + b + c
             P = 4' + 3' + 5'
             P = 12'

Example: Find the perimeter of this triangle.

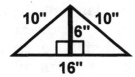

Solution: P = a + b + c
             P = 16" + 10" + 10"
             P = 36"

Key Point

The AREA of a triangle is equal to $\frac{1}{2}$ the base times the height. It is written as:

$$A = \frac{1}{2}b \times h$$

The height of the triangle will be the side of the figure only if the side forms a right angle (90°) with the base. A right angle is indicated as ⌐ in a drawing.

Example: Find the area of this triangle.

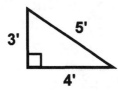

Solution: $A = \frac{1}{2}$ b x h

             $A = (\frac{1}{2})(4')(3')$

             A = (2')(3')
             A = 6 sq. ft.

Example: Find the area of this triangle.

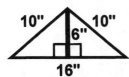

Solution: $A = \frac{1}{2}$ b x h

             $A = (\frac{1}{2})(16")(6")$

             A = (8")(6")
             A = 48 sq. in.

95

Section 5: The Pythagorean Theorem

Key Point

One of the formulas you'll be given is the Pythagorean Theorem, but before we discuss this formula, we must first introduce the arithmetic operation of taking a square root. The square root of a number is a number that when multiplied by itself will give us the original number.

Square Root is denoted by the symbol $\sqrt{}$.

$$\sqrt{9} = 3 \qquad\qquad \sqrt{81} = 9$$

Key Point

The Pythagorean Theorem states that if we square the length of both sides of a right triangle, and then take the square root of their sum, the result will be the length of the hypotenuse

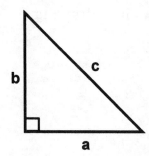

In the figure at the left, **a** and **b** are the sides of a right triangle. **c** is the hypotenuse. The relationship expressed in the Pythagorean Theorem can be expressed by the formula:

$c^2 = a^2 + b^2$, or

$$c = \sqrt{a^2 + b^2}$$

Example: Find the length of the hypotenuse of a right triangle whose sides are 3 feet and 4 feet.

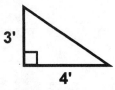

Solution: a = 3 and b = 4

$C^2 = a^2 + b^2 = 3^2 + 4^2$

$$C = \sqrt{3^2 + 4^2} = \sqrt{9 + 16} = \sqrt{25} = 5 \text{ feet}$$

Example: Find the length of the hypotenuse of a right triangle whose sides are 5 inches and 9 inches.

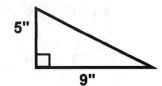

Solution: a = 5 and b = 9

$$c^2 = 5^2 + 9^2$$

$$c = \sqrt{5^2 + 9^2} = \sqrt{25 + 81} = \sqrt{106} = 10.2956 \text{ inches}$$

Key Point

With the addition of the calculator section on the test this is the type of question you will now see. On the old test questions on the Pythagorean Theorem always had answers that were whole numbers. This is not the case on the current series of tests.

Section 6: Similar Triangles

Key Point

Two Triangles are similar if their corresponding angles are equal and their corresponding sides are proportional.

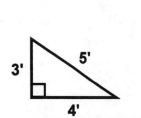

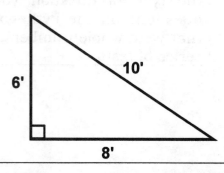

Key Point

If we can show that the corresponding angles (angles that are in the same position in each triangle) of one triangle are equal to the corresponding angles of a second triangle, the triangles are similar and the corresponding sides (sides that are in the same position in each triangle) will be proportional. In the example below, we have 2 right angles. The 2 angles formed by intersecting lines are equal. That means the third angles of the triangles are equal. The corresponding sides are the ones opposite the equal angles.

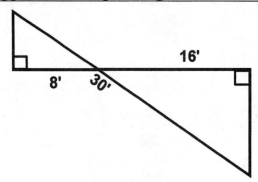

Example: If the horizontal line is cut into 8 foot and 16 foot segments by the 30 foot diagonal, what is the length of the hypotenuse of the smaller triangle?

Solution:

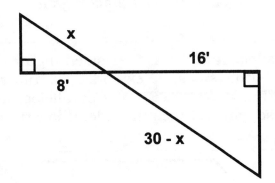

$$\frac{8}{x} = \frac{16}{30 - x}$$

16x = 240 – 8x
24x = 240
x = 10

Example: If ∠a = 70º, and ∠c is a right angle, what is the measurement of ∠e?

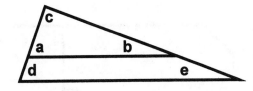

Solution: The triangle with angles c, a, and b, is similar to the triangle with angles c, d, and f.

∠a = ∠d and ∠b = ∠e .

Since we know that there are 180º in a triangle, and the sum of ∠a and ∠c = 160º, then ∠b = 20º.

Since ∠b = ∠e, then ∠e also = 20º.

Section 7: The Parallelogram

Key Point

A parallelogram is a four-sided figure with opposite sides parallel (the sides, if extended, would never intersect) and equal. The **PERIMETER** of a parallelogram is equal to 2 times the base, plus 2 times the side. It is written as:

P = 2b + 2c

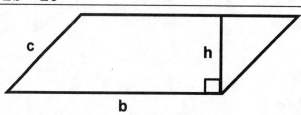

Let us apply this formula to the following examples.

Example: Find the perimeter of a parallelogram whose parallel sides are 10' and 5'.

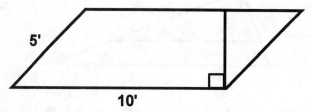

Solution: P = 2b + 2c
            P = 2(10') + 2(5')
            P = 20' + 10'
            P = 30'

Example: Find the perimeter of the parallelogram in the figure below.

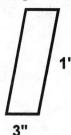

Solution: P = 2b + 2c
            P = 2(3") + 2(1')
            P = 2(3") + 2(12")
            P = 6" + 24"
            P = 30"

100

You must work in a common unit, which is why we change from 1' to 12" in the third step.

Key Point

The **AREA** of a parallelogram is equal to base times height. The formula differs from that of the rectangle only in that the height of the parallelogram is not necessarily the side of the figure.

$A = b \times h$

Example: Find the area of this figure if the height is equal to 4'.

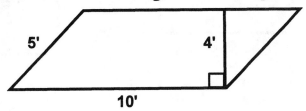

Solution: $A = b \times h$
$A = (10')(4')$
$A = 40$ sq. ft.

Example: Find the area of the figure below if the height is 10 inches.

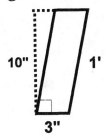

Solution: $A = b \times h$
$A = (3")(10")$
$A = 30$ sq. in.

Perimeter and Area of Squares, Rectangles, Triangle, and Parallelogram Problems

The following 3 examples refer to the four figures below. Figure (A) is a rectangle, figure (B) is a square, figure (C) is a parallelogram, and figure (D) is a right triangle. The solutions and explanations follow on the next page.

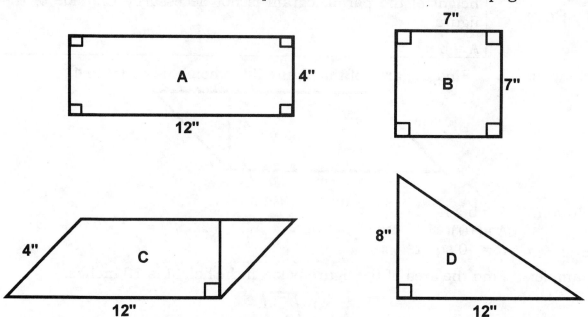

1. **In which two figures are the perimeters equal?**

   (1) (A) and (C)
   (2) (A) and (D)
   (3) (B) and (C)
   (4) There are no 2 perimeters that are equal
   (5) Not enough information is given.

2. **In which two figures are the areas equal?**

   (1) (A) and (C)
   (2) (A) and (D)
   (3) (B) and (C)
   (4) There are no 2 areas that are equal
   (5) Not enough information is given.

3. **Which figure has both the largest perimeter and largest area?**

   (1) (A)
   (2) (B)
   (3) (C)
   (4) (D)
   (5) No figure has both the largest perimeter and largest area..

**Perimeter and Area of Squares, Rectangles,
Triangle, and Parallelogram Solutions**

1.   1)   By simple inspection, this is obvious. Perimeter is equal to the sum of the sides, and (A) and (C) are both 4-sided figures with sides of equal length.

With some inspection, you can conclude that the perimeter of figure (D) is greater than the rectangle or parallelogram. The perimeter of figures (A) and (C) is equal to 8 + 12 + 12. The perimeter of the triangle is equal to 8 + 12 + a third number that we know to be greater than 12 because the hypotenuse of a right triangle is greater in length than either of its sides.

From a time standpoint this is the best way to approach to problem. However, in order to reinforce the use of formulas, we'll work out a detailed solution below.

Figure (A) is a rectangle. The opposite sides are equal in length. So, there are two 4 inch sides and two 12 inch sides.

The perimeter of figure (A) is equal to 2(4″) + 2(12″) = 8″ + 24″ = 32″

Figure (B) is a square. All four sides of a square are equal in length. So, there are four 7 inch sides.

The perimeter of figure (B) is equal to 4(7″) = 28″

Figure (C) is a parallelogram. The opposite sides are equal in length. So, there are two 4 inch sides and two 12 inch sides.

The perimeter of figure (C) is equal to 2(4″) + 2(12″) = 8″ + 24″ = 32″

Figure (D) is a right triangle. To find the length of the hypotenuse, use the Pythagorean Theorem.

The length of the third side equals

$$\sqrt{(8)^2 + (12)^2} = \sqrt{64 + 144} = \sqrt{208} = 14.42$$

The perimeter of figure (D) is approximately equal to

8″ + 12″ + 14.42″ = 34.42″

The correct answer is (1).

2.   2)   By inspection, we see we have a rectangle that has a height of 4″ and a base of 12″. We also have a triangle with a twice the height (8″) and the same base measurement (12″). Since the area of a rectangle is **l x h** and the area of a triangle is $\frac{1}{2}$**l x h**, we see that the rectangle (figure A) and the triangle (figure D) have the same area.

As with the previous problem, this time-saving method is the best way to approach to problem. However, we will again provide a detailed solution to reinforce the use of formulas.

Figure (A) is a rectangle. Its vertical side (4″) is its height and its horizontal side (12″) is its length.

The area of figure (A) = **l x h,** or, 4″ x 12″ = 48 sq. in.

Figure (B) is a square. All four sides of a square are equal in length. So, the height and the length are both 7″.

The area figure (B) is equal to 7″ x 7″ = 49 sq. in.

Figure (C) is a parallelogram. We don't know the height of the figure, but the perpendicular distance between the two 12″ sides must be less than slanted side (4″). So, we know that the area of figure (C) must be less than the area of figure (A).

Figure (D) is a right triangle, so its height is equal to the length of its perpendicular side (8″). The length of the triangle is 12″.

The area of figure (D) = $\frac{1}{2}$**l x h** or, $\frac{1}{2}$ x 8″ x 12″ = 48 sq. in.

The correct answer is (2).

3.　　5)　　We know from the previous two solutions that the square is larger in area (49 sq. ″) than either the rectangle or triangle (48 sq. ″), but it's smaller in perimeter (28″) than the rectangle (32″). Therefore none of the figures has both the largest area and perimeter.

The correct answer is (5).

Key Point

> The last four-sided figure we will look at is the trapezoid. A trapezoid is a four-sided figure having only two sides parallel. The parallel sides are called the **bases**. They are labeled **b** and **b'**. The **PERIMETER** of the trapezoid is equal to the sum of the four sides.
>
> P = a + b + c + b'

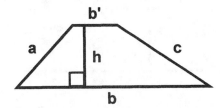

Let us apply this formula to the following problems.

Example: Find the perimeter of a trapezoid whose bases are 9" and 3" and whose sides are 4" and 5".

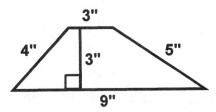

Solution: P = a + b + c + b'
P = 4" + 3" + 5" + 9"
P = 21"

Example: Find the perimeter of the trapezoid below.

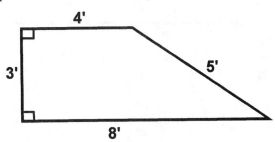

Solution: P = a + b + c + b'
P = 3' + 4' + 5' + 8'
P = 20'

Key Point

The **AREA** of a trapezoid is equal to the **average** of the bases times the height.

$$A = \frac{b + b'}{2} \bullet h$$

Example: Find the area of the trapezoid in figure below.

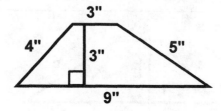

Solution: $A = \dfrac{b + b'}{2} \bullet h$

$A = \dfrac{3'' + 9''}{2} \bullet 3''$

$A = \dfrac{12''}{2} \bullet 3''$

$A = 6'' \bullet 3''$

$A = 18$ sq. in.

Example: Find the area of the trapezoid in the figure below.

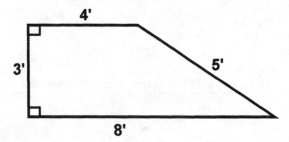

Solution: $A = \dfrac{b + b'}{2} \bullet h$

$A = \dfrac{4' + 8'}{2} \bullet 3$

$A = \dfrac{12'}{2} \bullet 3'$

$A = 6' \bullet 3'$

$A = 18$ sq. ft.

Section 9: Perimeter and Area Review

Let us review the figures we have studied so far by placing them next to each other. The appropriate formulas are under each diagram.

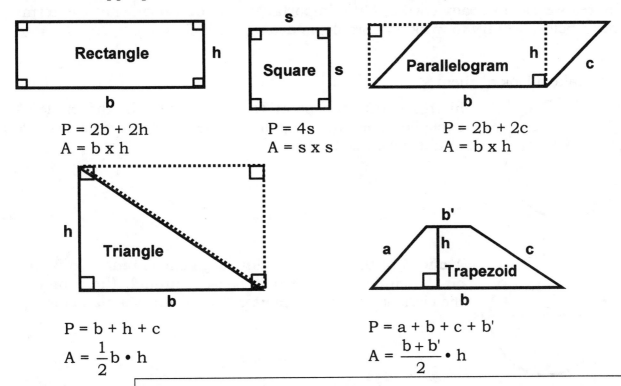

P = 2b + 2h
A = b x h

P = 4s
A = s x s

P = 2b + 2c
A = b x h

$P = b + h + c$

$A = \frac{1}{2}b \cdot h$

$P = a + b + c + b'$

$A = \frac{b + b'}{2} \cdot h$

Key Point

The dotted lines are used to show you that it is logical that the area of a parallelogram is the same as a rectangle and that the area of a triangle is half that of a rectangle. Notice that all area formulas involve a product of base times height, with the trapezoid using an average of its two bases.

Section 10: The Circle

A circle is a collection of points equidistant from a given point called the center. Because of its shape, the terms and formulas used are different from the figures we have already covered. It is important that you know what the terms mean before you try to work any problems dealing with circles.

Let's take a look at them.

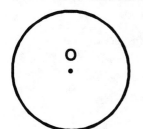

In this figure, the point O is the center of the circle. A circle is named by the point in the center. In this case we would say circle O.

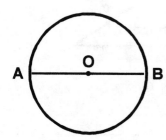

A **chord** of the circle is a line segment joining two points on the circle. A chord that passes through the center of the circle is called a **diameter**. Chord AB is a diameter.

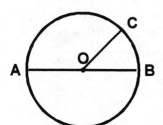

A line segment that joins the center with any point of the circle is called a **radius**. OA, OB and OC are all radii of the circle.

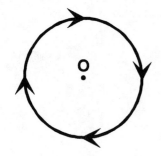

The **circumference** of a circle is equivalent to the perimeter of a rectangle. It is the distance around a circle.

A constant used in the formula for finding the circumference of a circle is written like this Π. It is equal to 3.14 or $\frac{22}{7}$. You can use either form, decimal or fraction. Use the one that matches the other numbers you are using in the problem.

Key Point

The formula for finding the **circumference** of a circle is:

C = 2Πr

Key Point

The formula for finding the **area** of a circle is Π times the radius squared (r x r).

A= Πr²

Example: Find the circumference and area of the circle below.

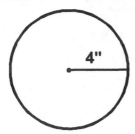

4"

Solution:

C = 2Πr	A = Πr²
C = 2Π(4")	A = Π(4")(4")
C = 2 (4") Π	A = (4")(4")(Π)
C = 8Π" *** See Note**	A = 16Π sq. in. *** See Note**
C = 8($\frac{22}{7}$)"	A = 16(3.14) sq. in.
C = $\frac{176}{7}$ inches	A = 50.24 sq. in.
C = 25 $\frac{1}{7}$ inches	

*NOTE: On the examination, check the form of the multiple-choice answers. You may not have to take your answer beyond this point.

Example: Find the area of the shaded region below.

Solution:

NOTE: If the diameter is 4', then the radius is half as large or 2'.
             Area of the square = s x s

$$A = 4' \times 4'$$

$$A = 16 \text{ sq. ft.}$$

Area of the circle = Πr^2

$$A = \Pi(2')(2')$$

$$A = 4\Pi \text{ sq. ft.}$$

Area of shaded region = Area of square - area of circle

Area of shaded region = $(16 - 4\Pi)$ sq. ft.

Numerical Relationships and Geometric Formulas Problems

1.   **If 6" square tile costs 45 cents each, how much is this by the square foot?**

2.   **If Jennifer drinks 6 eight-ounce glasses of water each day, how many gallons of water would she drink in a week?**

3.   **Twice a year, Neil donates a pint of blood at his company's blood drive. He will get a certificate from the blood bank when he donates a total of one gallon. How long will this take?**

110

Find the perimeter and area of the following figures:

4.

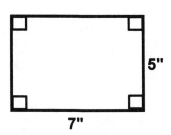

5"

7"

RECTANGLE

5.

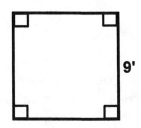

9'

SQUARE

6.

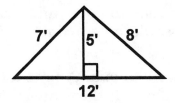

7' 5' 8'

12'

TRIANGLE

7.

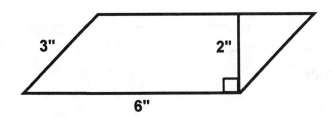

3" 2"

6"

PARALLELOGRAM

8.

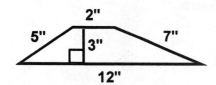

TRAPEZOID

9.

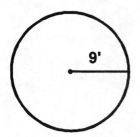

CIRCLE

10. If there are 43,560 square feet in an acre, which lot is larger, one that is a quarter acre or one that measures 100 feet along the front and is 120 feet deep?

Numerical Relationships and Geometric Formulas Solutions

1. If the tile is 6 inches on a side, it would take 4 tiles to cover an area 12 inches by 12 inches. Therefore, 1 square foot would cost 4 times .45 or $1.80.

2. There are 16 oz. in a pint, so six 8 oz. glasses would equal 3 pints a day. Over a week, this would equal 7 times 3 or 21 pints. If there are 2 pints to a quart and 4 quarts to a gallon, then there are 8 pints in a gallon. By dividing 8 into 21, we see that 21 pints is $2\frac{5}{8}$ gallons.

3. As we saw in the previous problem, there are 8 pints in a gallon. Neil starts by donating one pint, so has 7 more to donate. If he donates twice a year, it will take him $3\frac{1}{2}$ more years to donate a total of one gallon.

4. P = 2(7") + 2(5") = 14" + 10" = 24"

   A = (7")(5") = 35 sq. in.

5. P = 4(9') = 36'

   A = (9')(9') = 81 sq. ft.

6. P = 12' + 8' + 7' = 27'

   A = $\frac{1}{2}$(12')(5') = (6')(5') = 30 sq. ft.

7. $P = 2(6") + 2(3") = 12" + 6" = 18"$

   $A = (6")(2") = 12$ sq. in.

8. $P = 5" + 2" + 7" + 12" = 26"$

   $A = \dfrac{2" + 12"}{2} \bullet 2" = 7" \bullet 2" = 14$ sq. in.

9. $C = 2\Pi r = 2\Pi (9') = 18\Pi'$

   $A = \Pi r^2 = \Pi(9')(9') = 81\Pi$ sq. ft.

10. First, we have to find the size of the second lot. The area is 100' times 120' or 12,000 square feet. The easiest way to compare the two lots, is to multiply both by 4. In the first case, we will get one acre. In the second, we will get 48,000 square feet, which is more than an acre. The second lot, therefore, is the larger one.

Key Point

VOLUME is the size enclosed by a 3-dimensional figure.

We will now take a look at the formulas used to find the volume of a few geometric figures. The first is the rectangular solid.

RECTANGULAR SOLID

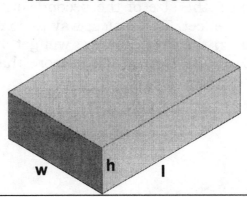

Key Point

The volume of a rectangular solid is equal to length times width, times height. The formula is written as:

V = l x w x h

The unit of measure for volume is cubic units.

Example: Find the volume of the figure below.

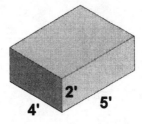

Solution: V = l x w x h

V = 5' x 4' x 2'

V = 40 cubic feet

CUBE

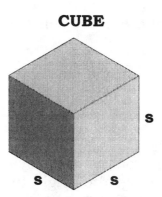

Key Point

> The second figure we will consider is the cube. It is the three dimensional counterpart to the square. The volume of a cube is equal to side times side times side or side-cubed. The formula is written as:
>
> $V = s \times s \times s$ OR $V = s^3$

Example: What is the volume of the cube below?

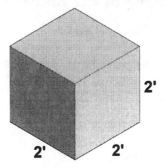

Solution: $V = s \times s \times s$ or s^3

$V = 2' \times 2' \times 2'$

$V = 8$ cu. Ft.

CYLINDER

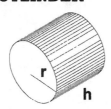

Key Point

> The last geometric solid we will discuss is the cylinder. The volume of a cylinder is equal to Π times radius-squared times height. An easy way to remember this is that the volume is equal to the area of the circular top times the height. The formula is written as:
>
> $V = \Pi r^2 h$

Example: Find the volume of the figure below.

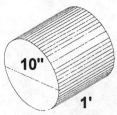

Solution: In order to find the volume of the cylinder, you must first find the radius. Second, change the height to inches. If the diameter is 10", then the radius is 5". The height is 1' or 12", so:

V = Πr²h

V = Π(5")(5")(12")

V = 300Π cubic inches.

Volume Problems

DIRECTIONS: **Find the volume of the following figures.**

1.   **A cube with a side equal to 9 inches.**

2.   **A cylinder whose radius is one foot and whose height is 6 inches.**

3.   **A rectangular solid whose length is 7 ft., width is 8 ft. and height is 6 ft.**

4.   **A cube with a side of $\frac{1}{2}$ a foot.**

5.   **A cylinder, given the area of its top as 16 Π sq. inches and its height is 9 inches.**

Volume Solutions

1.   V = 9" x 9" x 9"

     V = 729 cu. in

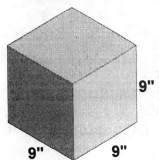

9"   9"   9"

116

2. $V = \Pi (12") (12") (6")$ 864Π cu. in.

or $V = \Pi(1')(1')(\frac{1'}{2}) = \frac{1}{2}\Pi$ cu. ft.

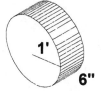

3. $V = 7' \times 8' \times 6'$

     $V = 336$ cu. ft.

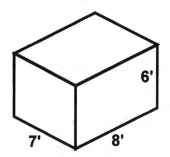

4. $V = \frac{1'}{2} \times \frac{1'}{2} \times \frac{1'}{2} = \frac{1}{8}$ cu. ft.

or $V = 6" \times 6" \times 6" = 216$ cu. in.

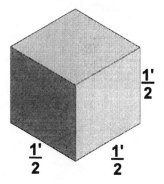

5. $V = \Pi r^2 h$

     $V = $ Area $\times$ h

     $V = 16\Pi" \times 9"$

     $V = 144\Pi$ cu. in.

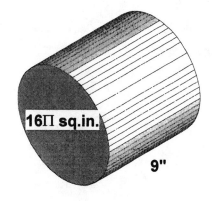

16Π sq.in.

9"

117

Section One: Algebraic Expressions

The expression (5)(4) or 5 * 4 represents the product of 5 and 4 which we know equals 20.

Key Point

The expression (5)(X) or 5 * X represents the product of 5 and some number. X is a **Variable**. When something is variable, it is unknown. In this case, it's the value of X that can unknown.

If X = 3, then 5 * X = 5 * 3 = 15

If X = 7, then 5 * X = 5 * 7 = 35

Key Point

The product of 5 * X can also be written in its most common form 5X. 5X is called an **Algebraic Term**. With our introduction of algebraic terms, we will no longer use "x" as a symbol of multiplication. X will be a variable.

3X + 7 is an **Algebraic Expression**. It's made up of one, or more, algebraic terms. Its value is a function of X. The number 3 is called the **Coefficient** of X. 7 is a **Constant**.

In the expression Y = 3X + 7, Y is a **Function** of X; its value is dependent on the value of X.

As we substitute different values for X, we generate new values for Y.

If X = 1, then Y = 3(1) + 7 = 10

If X = 7, then Y = 3(7) + 7 = 28

Example: If Y = X + 12, what is the value of Y, when X = 3?

1) 3
2) 12
3) 13
4) 15
5) 36

Solution: Substituting 3 in for X, we get Y = 3 + 12 = 15

4) is the correct answer.

Example: If $Y = \dfrac{2X + 7}{3X}$ what does Y equal when X = 3?

Solution: $Y = \dfrac{2(3) + 7}{3(3)} = \dfrac{6 + 7}{9} = \dfrac{13}{9} = 1\,\dfrac{4}{9}$

Example: In our equation Y = 3X + 7, what does X equal when Y = 49?

Solution: Y = 3X + 7

49 = 3X + 7

I don't know what 3X equals, but if I add 7 to it, I get 49. So, 3X must be = 42.

42 = 3X

I'm thinking of a number such that when I multiply it by 3, I get 42.

What's the number? Take a guess.

3 * 13 = 39 too small

3 * 14 = 42

14 = X

Now let's find out how to solve this problem without guessing.

Section Two: Equations

14 + 3 = 17 is an equation. It has terms on both sides of an equal sign and, in fact, the terms must be equal in value. If we were to add something to one side of the equation, we would have to add an equal amount to the other side.

$$14 + 3 + 4 = 17 + 4$$

If we were to take something away from one side of the equation, we would have to take an equal amount away from the other side.

$$14 + 3 - 2 = 17 + 4 - 2$$

Key Point

An equation that has an algebraic term on at least one side of the equal sign is an Algebraic Equation.

For Example: X + 5 = 50

X is an unknown. However, we're looking for the value of X that will make both sides of the equation equal. What number plus 5 equals 50? In this case, you might be able to figure this out in your head. The number is 45. In general, though, how would we solve an algebraic equation?

Key Point

Remember that your algebraic equation is a balanced scale. If you add something to one side, you must add the same value to the other side. If you multiply one side by a number, you must multiply the other side by the same number. The same goes for subtraction and division.

So, starting with:

$$\begin{array}{r} X + 5 = 50 \\ -5 = -5 \\ \hline X = 45 \end{array}$$

Example: Solve for X, if 2X + 9 = 15

Solution:

$$\begin{array}{r} 2X + 9 = 15 \\ -9 = -9 \\ \hline 2X = 6 \end{array}$$ First, subtract 9 from both sides of the equation

$$\frac{2X}{2} = \frac{6}{2}$$ Next, divide both sides by 2

$$X = 3$$

Example: If 3X – 12 = 48, what is X equal to?

    1) 12
    2) 16
    3) 20
    4) 36
    5) 60

Solution:

$$3X - 12 = 48$$
$$\underline{+12 = +12}$$ First, add 12 to both sides of the equation
$$3X = 60$$

$$\frac{3X}{3} = \frac{60}{3}$$ Next, divide both sides by 3

$$X = 20$$

The correct answer is 3).

Example: For what value of X is $\frac{2}{3}X - 5 = 13$ a true statement?

    1) $5\frac{1}{3}$
    2) 8
    3) 12
    4) 18
    5) 27

Solution: $\frac{2}{3}X - 5 = 13$ First, add 5 to both sides of the equation

$$\frac{2}{3}X - 5 + 5 = 13 + 5$$

$$\frac{2}{3}X = 18$$ Next, multiply both sides by 3/2

$$\frac{3}{2} \cdot \frac{2}{3}X = 18 \cdot \frac{3}{2}$$

$$X = 27$$

Key Point

> You could also do this in two steps by first multiplying both sides by 3 and then dividing both sides by 2.

The correct answer is 5).

Algebra Problems

1. **If 15 = X − 9, what does X equal?**

    1) 6
    2) 15
    3) 24
    4) 9
    5) 10

2. **If 3X + 4 = 16, what does X equal?**

    1) 12
    2) 20
    3) 3
    4) 4
    5) $\dfrac{20}{3}$

3. **If 5X − 30 = 2X, what does X equal?**

    1) 10
    2) 3
    3) 6
    4) $\dfrac{30}{7}$
    5) 15

4. **If 9X − 8 = 2X + 6, what does X equal?**

    1) 3
    2) $\dfrac{8}{9}$
    3) $\dfrac{14}{9}$
    4) 2
    5) $\dfrac{6}{7}$

5. **If X + 12 = 5X − 12, what does X equal?**

    1) 6
    2) 0
    3) 4
    4) 3
    5) 2

6. If $\frac{3}{4}X + 8 = 20$, what does X equal?

       1) 12
       2) 9
       3) 28
       4) 21
       5) 16

7. If $\frac{5x + 3}{4} = 7$, what does X equal?

       1) 2
       2) $\frac{4}{5}$
       3) 25
       4) 5
       5) $\frac{12}{5}$

8. If $\frac{2}{3}X + 1 = \frac{1}{3}X + 5$, what does X equal?

       1) 12
       2) 9
       3) 28
       4) 21
       5) 16

9. After a busy night at the restaurant, the waiters and waitresses count up their tips. Tom earned twice as Don, and Jennie earned as much as the two of them combined. If the total tips for the three of them came to $270.00, how much did Tom earn?

       1) $30
       2) $45
       3) $90
       4) $115
       5) $135

10. **Mike and Karen live 55 miles apart. They hop in their cars and drive towards each other, but Mike travels faster than Karen, and when the two meet, he ends up traveling 5 miles farther than Karen. How far did Mike travel before the two met?**

        1) 20
        2) 25
        3) 30
        4) 35
        5) 40

Algebra Solutions

1.
$$15 = X - 9$$
$$\underline{+9 = \quad +9}$$
$$24 = X$$

The correct answer is 3).

2.
$$3X + 4 = 16$$
$$\underline{\quad -4 = -4}$$
$$3X = 12$$

$$\frac{3X}{3} = \frac{12}{3}$$

$$X = 4$$

The correct answer is 4).

3.
$$5X - 30 = 2X$$
$$\underline{+30 = +30}$$
$$5X = 2X + 30$$

$$5X = 2X - 30$$
$$\underline{-2X = -2X}$$
$$3X = +30$$

$$\frac{3X}{3} = \frac{30}{3} \;, \; X = 10$$

The correct answer is 1).

4.

$$9X - 8 = 2X + 6$$
$$\underline{+ 8 = + 8}$$
$$9X = 2X + 14$$

$$9X = 2X + 14$$
$$\underline{- 2X = - 2X}$$
$$7X = + 14$$

$$\frac{7X}{7} = \frac{14}{7}, \quad X = 2$$

The correct answer is 4).

5.

$$X + 12 = 5X - 12$$
$$\underline{+ 12 = + 12}$$
$$X + 24 = 5X$$

$$X + 24 = 5X$$
$$\underline{- X = - X}$$
$$24 = 4X$$

$$\frac{24}{4} = \frac{4X}{4}, \quad X = 6$$

The correct answer is 1).

6.

$$\frac{3}{4}X + 8 = 20$$

$$\underline{- 8 = - 8}$$

$$\frac{3}{4}X = 12$$

$$\frac{4}{3} \bullet \frac{3}{4}X = 12 \bullet \frac{4}{3}$$

$$X = 16$$

The correct answer is 5).

7.

$$\frac{5X+3}{4}=7$$

$$4 \cdot \frac{5X+3}{4}=7 \cdot 4$$

$$5X+3=28$$

$$5X+3=28$$
$$\underline{-3=-3}$$
$$5X=25$$

$$\frac{5X}{5}=\frac{25}{5}=5$$

The correct answer is 4).

8.

$$\frac{2}{3}X+1=\frac{1}{3}X+5$$

$$\underline{-1=-1}$$
$$\frac{2}{3}X=\frac{1}{3}X+4$$

$$\frac{2}{3}X=\frac{1}{3}X+4$$

$$\underline{-\frac{1}{3}X=-\frac{1}{3}X}$$
$$\frac{1}{3}X=4$$

$$3 \cdot \frac{1}{3}X=3 \cdot 4=12$$

The correct answer is 1).

126

9. Let X represent amount of tips that Don earned. Since Tom earned twice as much as Don, we will let 2X represent Tom's tips. Jennie earned as much as Tom and Don combined, so that would be represented as X + 2X, or 3X. We also know that the sum of the tips equals $270. So,

X + 2X + 3X = 270

6X = 270

$$\frac{6x}{6} = \frac{270}{6}$$

X = 45

Remember, though, that X represents Don's tips. The question wants us to find the amount Tom earned. That is equal to 2X, or $90.

The correct answer is 3).

10. Let X represent the distance that Karen traveled. Since Mike traveled 5 miles farther, his distance will be represented by X + 5. We also know that the total distance traveled by the both of them is the 55 miles that originally separated them. So,

X + X + 5 = 55

2X + 5 = 55

$$\underline{-5 = -5}$$

2X = 50

$$\frac{2x}{2} = \frac{50}{2}$$

X = 25

Remember, though, that X represents Karen's distance. The question asked how far Mike traveled. That is equal to X + 5, or 30 miles

The correct answer is 3).

Section One: Points in Space

Coordinate or Analytical Geometry deals with location and relationship of points found on a two-dimensional surface. This surface is divided into 4 parts by two intersecting perpendicular lines called axes. The horizontal line is the x-axis and the vertical line is the y-axis. They're like horizontal and vertical number lines. An ordered pair of numbers, where the first number is always the distance traveled along the x-axis and the second number is the distance traveled along the y-axis, defines every point in this space.

Key Point

The point where the x-axis and y-axis meet is called the origin. It is the point where measurement originates. You move "0" in an x direction and "0" in a y direction to get there. We represent the origin by the notation (0,0), where the first number is the value you move in an x direction and the second number is the value you move in a y direction. Moving to the right or moving up from the origin is considered a positive direction. Moving to the left or moving down from the origin is considered moving in a negative direction.

Key Point

On the GED mathematics test, one of the new alternate format questions uses a grid with circles for you to darken in specific points. In the examples in this section we will use this grid so that you get used to seeing it. In real life, a grid can be drawn without all of the circles

Example: Locate the point (5,2)

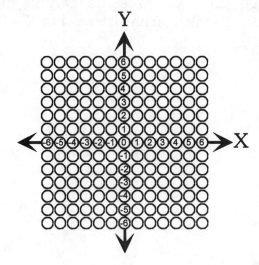

Solution: You would locate the point (5,2) by moving from the origin 5 spaces to the right and 2 spaces up.

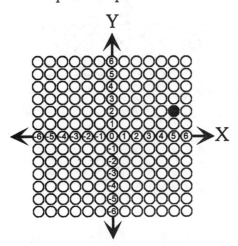

Example: Locate the point (-1,1)

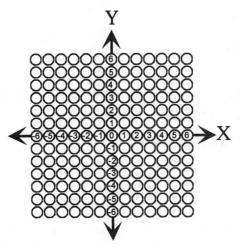

Solution: You would locate the point (-1,1) by moving from the origin 1 space to the left and 1 space up.

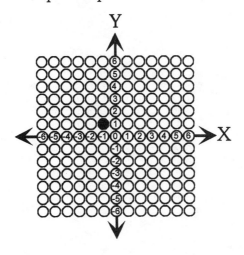

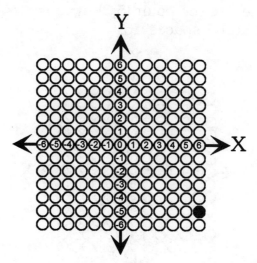

Example: What are the coordinates of the colored point?

      1) (6 , 5)
      2) (6 , -5)
      3) (-6 , 5)
      4) (-6 , -5)
      5) (-5 , 6)

Solution: In order to locate the point, from the origin, you move 6 units to right (+6) and 5 units down (-5). The point is (6, -5).

The correct answer is 2).

Section 2: Finding the Midpoint

Key Point

When you are given two points and want to find the midpoint, the steps to follow are straightforward. Simply add the two X coordinates together and divide by two to find the X coordinate of the midpoint. Then add the two Y coordinates together and divide by two to find the Y coordinate of the midpoint.

$$\frac{X_1 + X_2}{2} = X_m \qquad \frac{Y_1 + Y_2}{2} = Y_m$$

Example: Find the midpoint of the points (2,3) and (4,5).

Solution:

$$\frac{X_1 + X_2}{2} = X_m \qquad \frac{Y_1 + Y_2}{2} = Y_m$$

$$\frac{2+4}{2} = \frac{6}{2} = 3 = X_m \qquad \frac{3+5}{2} = \frac{8}{2} = 4 = Y_m$$

The midpoint equals (3,4)

Example: Grid in the midpoint of the points (-2,1) and (6,-1).

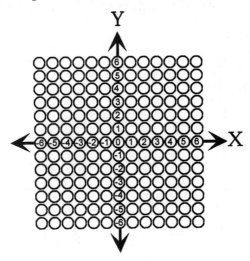

131

Solution: $$\frac{X_1 + X_2}{2} = X_m \qquad \frac{Y_1 + Y_2}{2} = Y_m$$

$$\frac{-2+6}{2} = \frac{4}{2} = 2 = X_m \qquad \frac{1+(-1)}{2} = \frac{0}{2} = 0 = Y_m$$

The midpoint equals (2,0)

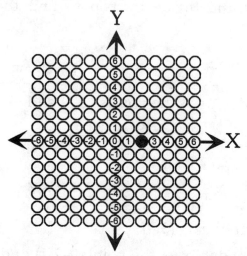

Example: Find the midpoint of the points (12,3) and (3,-5).

  1) (-9,-2)

  2) (7.5,1)

  3) (7.5,-1)

  4) (7.5,4)

  5) (7.5,-4)

Solution: $$\frac{X_1 + X_2}{2} = X_m \qquad \frac{Y_1 + Y_2}{2} = Y_m$$

$$\frac{12+3}{2} = \frac{15}{2} = 7.5 = X_m \qquad \frac{3+(-5)}{2} = \frac{-2}{2} = -1 = Y_m$$

The midpoint equals (7.5,-1)

The correct answer is 3).

Section 3: Graphing a Straight Line

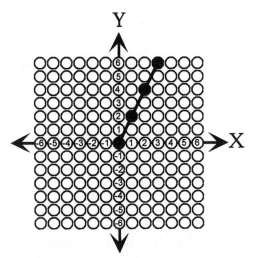

X	Y
0	0
1	2
2	4
3	6

When we connect any two points in our grid, and we extend our line, we notice that there is a constant relationship between the x and y coordinates of any point located on the line. In the example above, each time we move "1" in an X direction, we move "2" in a Y direction. This relationship between the change in Y over the change in X is called the slope of the line. The larger the positive slope, the steeper the angle of the line as you move from left to right. Lines with positive slopes rise from left to right. Lines with negative slopes fall from left to right.

Key Point

> In general, straight lines are represented by the equation:
>
> $y = mx + b$

The coefficient of x, m, is the slope of the line and the constant, b, is the y-intercept, which is the point at which the line crosses the y-axis. The y intercept has an x coordinate of zero. In our example above, the y-intercept is 0. The point (0,0) is called the origin.

If in the equation $y = mx + b$, we substitute 0 for x, we see that $y = b$. This shows us that, in fact, b is the value of y where the line crosses the y-axis.

If we wanted to find the equation of the line that passed through the points on our graph, we would first find the slope of the line. Remember that the slope is the ratio of the change in Y over the change in X.

The change in Y as you move from (0,0) to (1,2) is 2. The change in X is 1. The slope, m, is $\frac{2}{1}$ or 2.

So, we know that the equation is in the form $y = 2x + b$. Substituting any pair of values for x and y, we can solve for b.

$$y = mx + b$$
$$0 = 2(0) + b$$
$$0 = b$$

The equation is $y = 2x$

Example: What are the coordinates of the y-intercept for the line represented by $Y = \frac{2}{3}X - 7$

    1) (0,7)
    2) (2,3)
    3) (-7,0)
    4) (0,-7)
    5) (0,0)

Solution: In this example, the value of "b" is –7, so the y-coordinate of the y-intercept is –7. The x-coordinate of the y-intercept is always zero. So the y-intercept is (0,-7).

The correct answer is 4).

Example: What is the slope of a line that goes through the points (1,5) and (3,9).

Solution: The measurement in the change in y from (1,5) to (3,9) is the difference between the two y values. The difference between 9 and 5 is 4. The change in the x values is the difference between 3 and 1, or 2. The change in y (4) over the change in x (2) defines the slope of the line. The slope, "m" is equal to $\frac{4}{2}$, or 2.

Example: What is the equation of a line that goes through the points (1,5) and (3,9).

Solution: So, now we know that the equation that defines the line passing through (1,5) and (3,9) is of the form:

$$y = 2x + b$$

We also know that when x = 1, y = 5. So, we can substitute these values into the equation, and solve for b.

$$y = 2x + b$$

$$5 = 2(1) + b$$

$$5 = 2 + b$$

$$3 = b$$

The equation of the line that crosses through both (1,5) and (3,9) is:

$$y = 2x + 3$$

Example: What is the equation of a line that goes through the points (6,0) and (12,4).

Solution: The measurement in the change in y from (6,0) to (12,4) is the difference between the two y values. The difference between 4 and 0 is 4. The change in the x values is the difference between 12 and 6, or 6. The change in y (4) over the change in x (6) defines the slope of the line. The slope, "m" is equal to $\frac{4}{6}$, or $\frac{2}{3}$.

So, now we know that the equation that defines the line passing through (6,0) and (12,4) is of the form:

$$y = \frac{2}{3}x + b$$

We also know that when x = 6, y = 0. So, we can substitute these values into the equation, and solve for b.

$$y = \frac{2}{3}x + b$$

$$0 = \frac{2}{3}(6) + b$$

$$0 = 4 + b$$

You have to subtract 4 from 4 in order to get zero, so b must be -4

The equation of the line that crosses through both (6,0) and (12,4) is:

$$y = \frac{2}{3}x - 4$$

Example: Below are values of x and y along a straight line. Find the missing value in the table.

x	y
4	19
2	13
0	?

1) 7
2) -7
3) 5
4) -5
5) 0

135

Solution: The slope of the line (m) is equal to the change in y over the change in x. This is equal to:

$$\frac{(19 - 13)}{(4 - 2)} = \frac{6}{2} = 3$$

So, the equation of our line is equal to $y = 3x + b$. We know that $y = 13$ when $x = 2$ so we can substitute these values in our equation.

$y = 3x + b$

$13 = 3(2) + b$

$13 = 6 + b$

$7 = b$

Our equation is $y = 3x + 7$. When $x = 0$, $y = 7$.

The correct answer is 1).

Key Point

Section 4: Distance Between Two Points

The distance between two points is represented by the formula $\sqrt{(X_2 - X_1)^2 + (Y_2 - Y_1)^2}$, where the coordinates of the starting point are (X_1, Y_1) and the coordinates of the end point are (X_2, Y_2). By subtracting one X coordinate from the other, we are calculating the change in the X value, or the distance traveled in an X direction. By subtracting one Y coordinate from the other, we are calculating the change in the Y value, or the distance traveled in an Y direction.

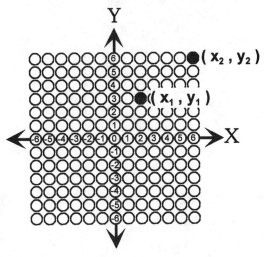

If you think of the distance traveled in the X direction as the base of a right triangle, and the distance traveled in the Y direction as the side of a right triangle, then the distance between the two points is the hypotenuse of the right triangle.

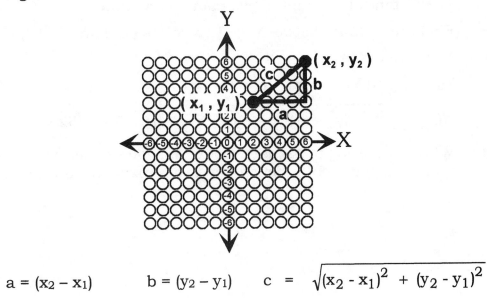

$$a = (x_2 - x_1) \qquad b = (y_2 - y_1) \qquad c = \sqrt{(x_2 - x_1)^2 + (y_2 - y_1)^2}$$

137

So, the formula:

$$\sqrt{(x_2 - x_1)^2 + (y_2 - y_1)^2}$$

is simply an application of the Pythagorean Theorem.

Example: Find the distance between the two points below.

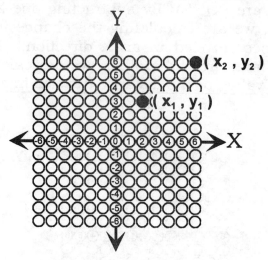

Solution: The coordinates of the first point are (2,3). The coordinates of the second point are (6,6)

$X_2 = 6$ and $X_1 = 2$; $X_2 - X_1 = 4$

$Y_2 = 6$ and $Y_1 = 3$; $Y_2 - Y_1 = 3$

$$\sqrt{(x_2 - x_1)^2 + (y_2 - y_1)^2} = \sqrt{4^2 + 3^2} = \sqrt{16 + 9} = \sqrt{25} = 5$$

The distance between the two points is 5 units.

Example: Find the distance between the two points (1,6) and (5,12)

Solution: $X_2 = 5$ and $X_1 = 1$; $X_2 - X_1 = 4$

$Y_2 = 12$ and $Y_1 = 6$; $Y_2 - Y_1 = 6$

$$\sqrt{(x_2 - x_1)^2 + (y_2 - y_1)^2} = \sqrt{4^2 + 6^2} = \sqrt{16 + 36} = \sqrt{52} = 7.21$$

When evaluating the square root of 52, you can use your calculator and the square root function key to approximate your answer.

Coordinate Geometry Problems

1. **Locate the point (6,-3) on the grid below.**

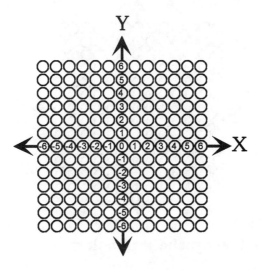

2. **Locate the point (-4,-5) on the grid below.**

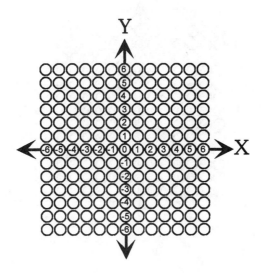

3. Locate the point (4,5) on the grid below.

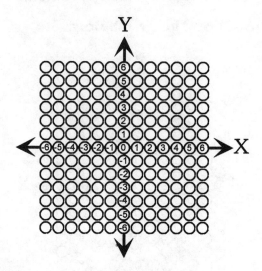

4. Locate the point (-6,2) on the grid below.

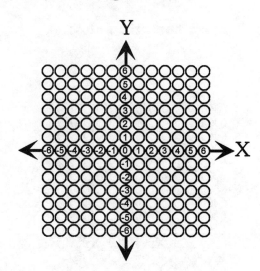

5. Locate the point (0,-3) on the grid below.

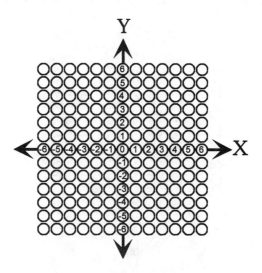

6. Locate the point (-2,0) on the grid below.

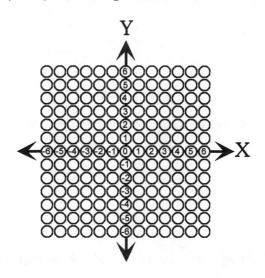

7. **Grid in the midpoint of the points (-3,4) and (-1,2).**

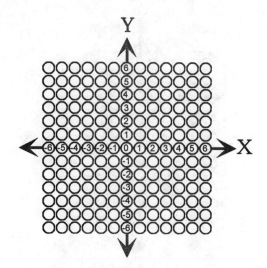

8. **What is the midpoint of the points (8,2) and (12,-8)?**

     1) (8,0)
     2) (-4,10)
     3) (-4,-6)
     4) (20,-6)
     5) (10,-3)

9. **What is the y-intercept of the line with and equation of y = 3x + 4?**

     1) (4,0)
     2) (0,3)
     3) (0,-3)
     4) (0,-4)
     5) (0,4)

10. **What is the y-intercept of the line with the equation of y = $\frac{1}{4}$?**

     1) (4,0)
     2) (0,4)
     3) (0,0)
     4) (0,$\frac{1}{4}$)
     5) ($\frac{1}{4}$,0)

11. Grid in the y-intercept of the equation y = -2x – 3.

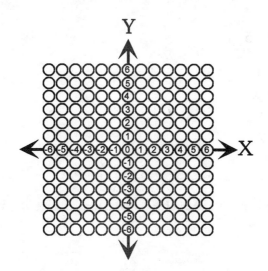

12. What is the slope of the line with the equation y = 2x – 3?

    1) 2
    2) 3
    3) –3
    4) –2
    5) $-\dfrac{2}{3}$

13. What is slope of the line with and equation of y = 3x + 4?

    1) 4
    2) -4
    3) -3
    4) $\dfrac{3}{4}$
    5) 3

14. What is the slope of the line with the equation of y = $\dfrac{1}{4}$?

    1) 4
    2) 0
    3) -4
    4) $\dfrac{1}{4}$
    5) $-\dfrac{1}{4}$

15. **What is the equation of a line that passes through the points (3,5) and (5,8)?**

 1) $y = 1.5x + .5$

 2) $y = 15x$

 3) $y = \dfrac{2}{3}x + \dfrac{1}{2}$

 4) $y = \dfrac{2}{3}x - \dfrac{1}{2}$

 5) $y = x - \dfrac{1}{2}$

16. **What is the length of a line that connects the points (3,5) and (5,8)?**

 1) 5

 2) 4

 3) 3.6

 4) 13

 5) 17.72

17. **What is the length of a line that connects the points (-2,2) and (4,10)?**

 1) 3

 2) 4

 3) 5

 4) 6

 5) 10

Coordinate Geometry Solutions

1. The point (6,-3) is located by moving in the X direction 6 spaces to the right and in the Y direction 3 spaces down.

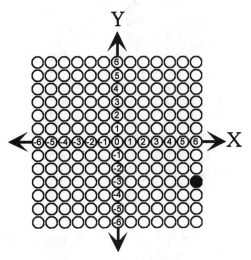

2. The point (-4,-5) is located by moving in the X direction 4 spaces to the left and in the Y direction 5 spaces down.

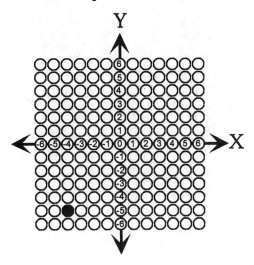

3. The point (4,5) is located by moving in the X direction 4 spaces to the right and in the Y direction 5 spaces up.

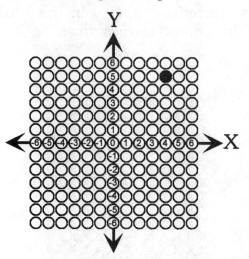

4. The point (-6,2) is located by moving in the X direction 6 spaces to the left and in the Y direction 2 spaces up.

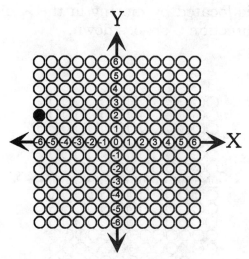

5.   The point (0,-3) is located by moving in the X direction 0 spaces and in the Y direction 3 spaces down.

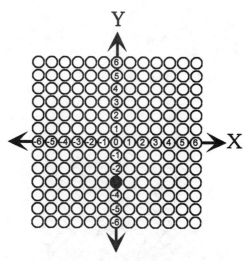

6.   The point (-2,0) is located by moving in the X direction 2 spaces to the left and in the Y direction 0 spaces.

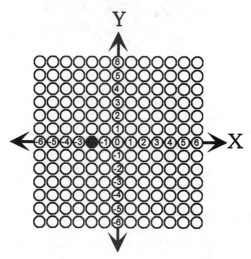

7.
$$\frac{X_1 + X_2}{2} = X_m \qquad\qquad \frac{Y_1 + Y_2}{2} = Y_m$$

$$\frac{(-3) + (-1)}{2} = \frac{-4}{2} = -2 = X_m \qquad\qquad \frac{4 + 2}{2} = \frac{6}{2} = 3 = Y_m$$

The midpoint equals (-2,3)

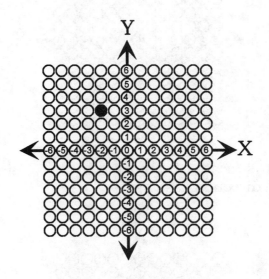

8.
$$\frac{X_1 + X_2}{2} = X_m \qquad\qquad \frac{Y_1 + Y_2}{2} = Y_m$$

$$\frac{8 + 12}{2} = \frac{20}{2} = 10 = X_m \qquad\qquad \frac{2 + (-8)}{2} = \frac{-6}{2} = -3 = Y_m$$

The midpoint equals (10,-3)

The correct response is 5).

9. To find the y-intercept you plug 0 into the equation for x.

y = 3(0) + 4 = 4

The y-intercept equals (0,4)

The correct response is 5).

10. A line with the equation $y = \frac{1}{4}$ has no slope. For any x, y equals $\frac{1}{4}$.

Therefore, when x = 0, y will equal $\frac{1}{4}$.

The correct response is 4).

11. The equation y = -2x – 3 has a y intercept of (0,-3).

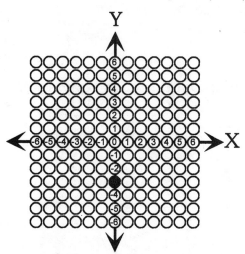

12. The slope of a line is equal to the coefficient of x in its equation.

    y = 2x – 3   The slope equals 2.

    The correct response is 1).

13. The slope of a line is equal to the coefficient of x in its equation.

    y = 3x + 4   The slope equals 3.

    The correct response is 5).

14. The slope of a line is equal to the coefficient of x in its equation.

    $y = 0x + \frac{1}{4}$ The slope equals 0.

    The correct response is 2).

15. The slope of the line (m) is equal to the change in y over the change in x. This is equal to:

$$\frac{(8-5)}{(5-3)} = \frac{3}{2} = 1.5$$

So, the equation of our line is equal to y = 1.5x + b. We know that y = 8 when x = 5 so we can substitute these values in our equation.

y = 1.5x + b

8 = 1.5(5) + b

8 = 7.5 + b

.5 = b

Our equation is y = 1.5x + .5

The correct answer is 1).

16. $X_2 = 5$ and $X_1 = 3$; $X_2 - X_1 = 2$

$Y_2 = 8$ and $Y_1 = 5$; $Y_2 - Y_1 = 3$

$$\sqrt{(x_2 - x_1)^2 + (y_2 - y_1)^2} = \sqrt{2^2 + 3^2} = \sqrt{4 + 9} = \sqrt{13} = 3.605$$

When evaluating the square root of 13, you can use your calculator and the square root function key to approximate your answer.

The correct answer is 3).

17. $X_2 = 4$ and $X_1 = -2$; $X_2 - X_1 = 6$

$Y_2 = 10$ and $Y_1 = 2$; $Y_2 - Y_1 = 8$

$$\sqrt{(x_2 - x_1)^2 + (y_2 - y_1)^2} = \sqrt{8^2 + 6^2} = \sqrt{64 + 36} = \sqrt{100} = 10$$

When evaluating the square root of 13, you can use your calculator and the square root function key to approximate your answer.

The correct answer is 5).

Section 1: The Circle Graph

Key Point

First type of graph we'll look at is a Circle Graph, also called a Pie Chart. It's used when you wish to show an entire quantity and the parts that make it up. Each part represents a percentage of the whole and the sum of the parts equals 100% of the circle or pie.

Below is a pie chart representing the percentages of different monthly expenses.

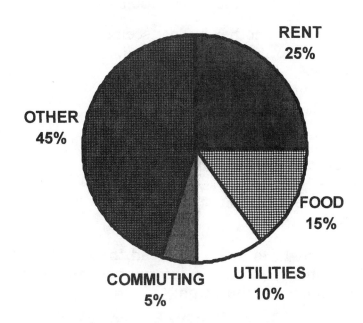

Example: What percent of the budget is allocated to maintaining the apartment?

1) 10%
2) 25%
3) 35%
4) 40%
5) 45%

Solution: Food, Commuting, and "Other" are not expenses associated with the apartment. Rent and Utilities are. The sum of these two equals 35%.

The correct answer is 3).

Example: What three items represent 50% of the monthly budget?

1) Rent, Food, & Utilities
2) Food, Utilities, & Commuting
3) Utilities, Commuting, & Other
4) Rent, Commuting, & Other
5) Rent, Utilities, & Commuting

Solution: The three items that add up to 50% are Rent (25%), Food (15%) and Utilities (10%).

The correct answer is 1).

Pie Chart Problems

The circle graph below is another budget scenario, but is a little more complex due to the amount of information represented.

rent - 25%
car payments - 10%
food - 15%
clothing - 10%
gas, etc. - 10%
medical - 5%
savings - 10%
entertainment - 15%

It is not enough, though, to be able to read these numbers off the graph. You will be asked questions about the information and will have to be able to interpret the data shown on the graph.

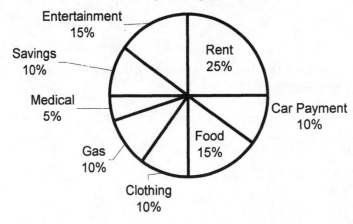

Mike's Monthly Budget - $4000

152

Try these questions based on the graph of Mike's budget.

1. **What percent of the budget is allocated for bills?**

    1) 25%
    2) 50%
    3) 75%
    4) 100%
    5) 85%

2. **What is the most expensive budget item?**

    1) rent
    2) food
    3) clothing
    4) medical
    5) entertainment

3. **There is as much money spent on entertainment each month as there is on which other item?**

    1) car
    2) clothing
    3) food
    4) rent
    5) gas, electric and phone

4. **What is the car payment each month?**

    1) $236
    2) $400
    3) $600
    4) $800
    5) $1000

5. **How much more is spent on rent than on medical expenses?**

    1) $200
    2) $400
    3) $600
    4) $800
    5) $1000

Pie Chart Solutions

1. The parts of the pie that could be considered bills would be:

Rent	25%
Car	10%
Food	15%
Clothing	10%
Gas, electric & phone	10%
Medical	5%
Total	75%

Bills are 75% of the budget, so "3" is the correct response.

In this example, it might have been easier to total the non-bill items and subtract their sum from 100%. You would have:

Savings	10%
Entertainment	15%
Total	25%

Non-Bills = 100%- 25% = 75%

2. At 25% of the total budget, rent represents the largest "piece of the pie".

The correct answer is 1).

3. Entertainment is 15% of the budget. The only other item that is of equal size is food.

The correct answer is 3).

4. The total monthly income is $4,000, of which 10% goes to the car payment.

    10% of $4,000 = $4,000 x .10 = $400

The correct answer is 2).

5. Rent is 25% of the budget; 25% of $4,000 equals $1000. Medical expenses are 5% of the budget; 5% of $4,000 = $200. The difference between rent and medical expenses equals $1000 - $50 = $800.

The correct answer is 4).

154

Section 2: The Bar Graph

> The second type of graph we'll examine is the bar chart. A bar graph is used for the quick comparison of data. Although it accomplishes this purpose, its drawback is that it is difficult to read accurately. Data is represented by solid bars - the longer the bar, the larger the quantity.

Points Scored by Week

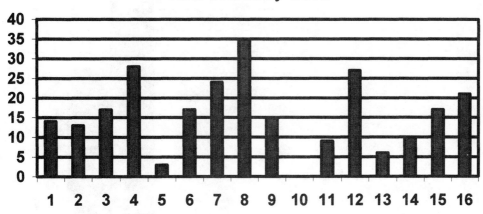

Example: What is mode score?

    1) 10
    2) 17
    3) 21
    4) 28
    5) 35

Solution: The mode is the number of points the team scored most often. In the 3rd, 6th, and 15th week, the team scored the same number of points. The value is somewhere between 15 and 20. It looks to be 17, which is the only answer between 15 and 20.

The correct answer is 2).

Example: In which week did the team score the most number of points?

    1) 1
    2) 4
    3) 8
    4) 12
    5) 16

Solution: The highest bar is in week 8.

The correct answer is 3).

Bar Chart Problems

The chart below contains partial statistics about New Hampshire Jobs by Industry.

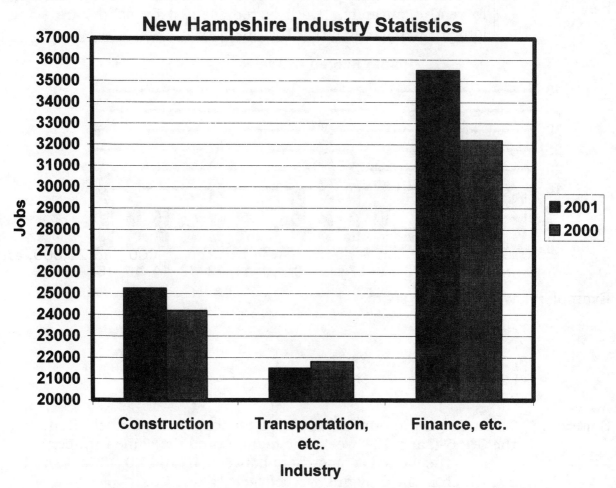

The following two questions refer to the graph above.

1. Which industries saw an increase in jobs from 2000 to 2001?

        1) Construction and Transportation
        2) Construction and Finance
        3) Transportation and Finance
        4) All three Industries
        5) None of the Industries

2. **Approximately, what was the total number of jobs in these three industries in 2001?**

       1) 77,750
       2) 79,250
       3) 80,000
       4) 80,500
       5) 82,250

Bar Chart Solutions

1. For each industry, the bar to the left represents 2001. The bar on the left is higher than the bar on the right (2000) for Construction and Finance.

The correct answer is 2).

2. Make sure that you sum only the values of the left-hand bars for each industry. Construction is about ¼ of the way between 25,000 and 26,000, Transportation is about ½ way between 21,000 and 22,000, and Finance is about ½ way between 35,000 and 36,000. We know our sum is going to be greater than the sum of the lower limits (25,000 + 21,000 + 35,000 = 81,000), but less than the sum of our upper limits (26,000 + 22,000 + 36,000 = 84,000). This eliminates 1)-4) as possible correct answers. This only leaves choice 5).

The correct answer is 5).

Key Point

The third and final type of graph covered in this section is the line graph. It is easier to read than the bar graph and highlights trends in the data being represented. Dots are placed at points that would be equivalent to the tip of the bar in the prior example. These dots are connected forming the line of the graph.

Line Graph Problems

The following questions refer to the graph below.

Kilowatt Hours Used

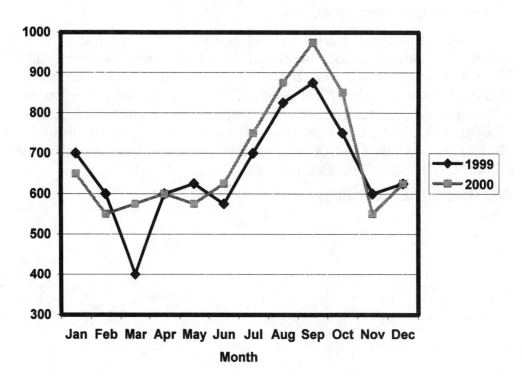

1. **For the year 2000, in what month did the highest electric use occur?**

   1) January
   2) March
   3) August
   4) September
   5) December

2. In how many months was electric usage greater in 1999 than in 2000?

  1) 2
  2) 3
  3) 4
  4) 5
  5) 6

3. During which two months was there no change in usage from 1999 to 2000?

  1) April & December
  2) January & February
  3) May & June
  4) November & December
  5) February & November

Line Graph Solutions

1. Drawing a straight line from the peak of the lighter-shaded line to the bottom of the graph. The line will cross the horizontal axis at "Sep".

The correct answer is 4).

2. The darker-shaded line represents 1999. It is above the lighter-shaded line on four occasions – January, February, May, and November.

The correct answer is 3).

3. There are two places on the graph where 1999 and 2000 appear to take on the same value – April and December. In fact, the square from the 2000 line covers the diamond from the 1999 line. These are the two months where there was no change in usage from one year to the next.

The correct answer is 1).

Section 4: Graphs and Chart Sample Problems

Problems 1-3 refer to the chart below.

Occupational Distribution Workers
Total Employment 25,594,794

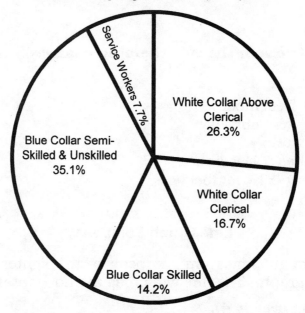

1. **The largest number of workers are employed in:**

   1) white collar above clerical
   2) white collar clerical
   3) blue collar skilled
   4) blue collar semi-skilled and unskilled
   5) service workers

2. **How many workers are employed as service workers?**

   1) 1,970,799
   2) 2,190,924
   3) 4,274,330
   4) 4,274,331
   5) Cannot be determined from the given information

3. **How many more people are employed as white collar clerical than as blue collar skilled?**

   1) 230,353
   2) 639,870
   3) 2,252,300
   4) 4,709,442
   5) 5,349,311.9

Problems 4-5 refer to the graph below.

Frequency of doctors' visits as it relates to education of family head.

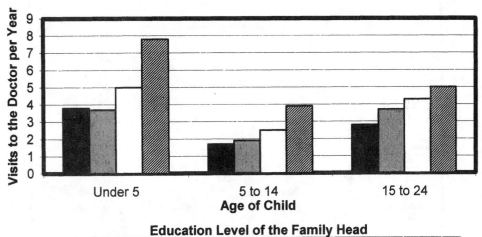

4. **In comparing the under 5 years age group with the 15-24 years age group, the group that shows little change is:**

       1) under 5 years of age
       2) under 5 years completed
       3) 5-8 years
       4) 9-12 years
       5) 13+ years

5) **The group that consistently gets less medical care is:**

       1) under 5 years
       2) 5-14 years
       3) 15-24 years
       4) 13+ years of education
       5) less than 5 years of education

Problems 6-8 refer to the figure below.

Men in Military Service

6. **During what period was there the greatest increase in total number of service men.**

   1) 1950-1951
   2) 1952-1953
   3) 1960-1961
   4) 1964-1965
   5) 1966-1967

7. **In 1948, how many men between the ages of 17 and 20 were in the service?**

   1) .5
   2) 50,000
   3) 500,000
   4) 5,000,000
   5) 5,000

8. **For the period from 1948-1967, the number of men in military service peaked during which year?**

   1) 1967
   2) 1966
   3) 1962
   4) 1953
   5) 1952

Graph and Chart Solutions

1. The largest piece of the pie is blue collar, semiskilled, and unskilled

    The correct answer is 4).

2. 7.7% of 25,594,794 = 25,594,794 x .077 = 1,970,799

    The correct answer is 1).

3. (16.7 - 14.2)% of 25,594,794 = 25,594,794 x .025 = 639,870

    The correct answer is 2).

4. The 5-8 years of education bars in both groups are both the same length.

    The correct answer is 3).

5. The bars for the 5-14 years group are smaller than the under 5 or the 15-24 years in each category.

    The correct answer is 2).

6. The graph is the steepest for the period from 1950-1951.

    The correct answer is 1).

7. .5 million or 500,000

    The correct answer is 3).

8. The highest point on the graphs at 1952.

    The correct answer is 5).

Questions 1 - 7 refer to the following situation.

**Lauren's gross monthly income is $1,600, but she is about to get a 25%
raise. The company, however, is going to withhold 25% of total income to
cover state and federal taxes, Social Security, unemployment insurance,
and medical benefits. Lauren is faced with a decision of either going into
an apartment on her own where the rent will be $\frac{1}{3}$ of her monthly take
home pay, or take an apartment with a friend where her share of the rent
would be 20% of her gross monthly pay. In the first case, she could walk
to work from the apartment. In the second case, it would cost her $10.00
a week for gas, $1.00 a day on tolls, and $150.00 more each year for car
insurance. Lauren's decision will be a financial, not an emotional one.**

1. What Is Lauren's net monthly income after her raise?

      1) $1,200
      2) $1,500
      3) $1,600
      4) $1,750
      5) $2,000

2.   What will her monthly rent be if she shares an apartment?

      1) $240
      2) $300
      3) $320
      4) $350
      5) $400

3.   What would her rent be if she decides to live on her own?

      1) $400
      2) $450
      3) $500
      4) $533
      5) $667

**4.   What will her savings on rent be if she decides to share an
apartment?**

      1) $100
      2) $150
      3) $175
      4) $200
      5) $267

5. If Lauren decides to live with her friend, she must commute to work. What is the total annual cost of her commute if she works 5 days a week, 50 weeks a year?

    1) $250
    2) $500
    3) $750
    4) $900
    5) $975

6. Since the real cost of the shared apartment is the rent plus commuting costs, what is the real annual expense of the second apartment?

    1) $4,800
    2) $5,300
    3) $5,450
    4) $5,550
    5) $5,700

7. Which apartment should Lauren choose and what will be the annual savings?

    1) Shared, $100
    2) Single, $150
    3) Single, $275
    4) Shared, $300
    5) Insufficient data is given to solve the problem

Questions 8 - 11 refer to the following situation.

Chris and Kathy want to paint their bedroom which measures 12 feet by 10 feet with an 8 foot ceiling. There are two windows each 1 yard square, a closet door 2' by 7', and the entrance door is 1 yd. by 7 ft. The paint they are going to use normally sells for $16.40 per gallon, but is on sale now for 25% off. The ad guarantees that one gallon will cover 300 square feet.

8. What is the total area of all the doors and windows?

    1) 23 sq. ft.
    2) 27 sq. ft.
    3) 39 sq. ft.
    4) 53 sq. ft.
    5) 60 sq. ft.

9. **What is the total surface area that they have to paint?**

        1) 123 sq. ft.
        2) 187 sq. ft.
        3) 299 sq. ft.
        4) 352 sq. ft.
        5) 427 sq. ft.

10. **If they must buy paint by the gallon, how much will they spend on paint?**

        1) $4.10
        2) $8.20
        3) $12.30
        4) $16.40
        5) $24.60

11. **How much did they save by buying the paint on sale?**

        1) $4.10
        2) $8.20
        3) $12.30
        4) $16.40
        5) $24.60

12. **Bob and Colleen just bought a new house and their first project is to put up a picket fence around the entire yard. If the fencing is sold in one-yard sections costing $2.10 each, and if the yard measures 99 feet by 150 feet, how much will the entire fence cost?**

        1) $174.30
        2) $348.60
        3) $522.90
        4) $1045.80
        5) $3465.00

13. **If cat food is on sale 3 cans for a dollar, how much would 20 cans cost?**

        1) $6.60
        2) $6.66
        3) $6.67
        4) $6.80
        5) $7.00

Questions 14 through 17 refer to the following situation.

In order to earn money while he is in college, Tommy works part time as a math tutor for $10/hr. and drives a truck for an auto parts store for $8/hr. He always works for the auto parts store from 3:00 to 5:00 Monday through Friday. The first full week in October, he tutors for 3 hours, the second week, just before mid-terms, he tutors 10 hours, the third week he does not tutor at all, but he works 8 hours on Saturday at the auto supply, and the last full week he tutors 5 hours.

14. **What is Tommy's average weekly wage tutoring?**

   1) $36
   2) $45
   3) $52
   4) $65
   5) $180

15. **What are his total wages for the month at the auto supply shop?**

   1) $80
   2) $320
   3) $384
   4) $400
   5) $480

16. **What is his average weekly wage considering both sources of income?**

   1) $56
   2) $125
   3) $141
   4) $152
   5) $165

17. **Which sequence correctly fists the weeks in order of most income to least income?**

   1) 2, 4, 1, 3
   2) 3, 2, 4, 1
   3) 2, 3, 1, 4
   4) 2, 4, 3, 1
   5) 2, 3, 4, 1

Questions 18 and 19 refer to the following situation.

Lois stops for gas. If she uses the oil company credit card she will save 2%. The price listed on the pump is $1.38 $\frac{9}{10}$ cents per gallon.

18. **If Lois pays with the oil company credit card, how much would 10 1/2 gallons cost?**

        1) $14.37
        2) $14.29
        3) $14.58
        4) $14.88
        5) Insufficient data is given to solve the problem

19. **How much did she save by paying the oil company credit card?**

        1) $.02
        2) $.03
        3) $.29
        4) $.39
        5) Insufficient data is given to solve the problem

20. **Jack wants to put baseboard molding in a room that measures 11' 4" long and 8 $\frac{1}{2}$ feet wide and has an entrance 3 feet wide. How much molding does he need?**

        1) 19 $\frac{5}{6}$ feet

        2) 35 $\frac{1}{10}$ feet

        3) 36 $\frac{2}{3}$ feet

        4) 39' 8"

        5) 76 $\frac{1}{3}$ feet

21. **If 100 eggs cost $9.00, what is the price per dozen?**

        1) $.09
        2) $.12
        3) $.90
        4) $1.08
        5) $1.20

Questions 22 - 26 refer to the following situation.

Of the 1000 students in a high school, 800 take at least one foreign language, but no one takes more than two. 400 take Spanish, 300 take French, 200 take Latin, and 100 take German.

22. **How many students are taking more than one foreign language?**

    1) none
    2) 100
    3) 200
    4) 300
    5) Insufficient data is given to solve the problem

23. **What percentage of all the students take Spanish?**

    1) 10%
    2) 20%
    3) 30%
    4) 40%
    5) 50%

24. **What percentage of all the students taking a foreign language are taking Spanish?**

    1) 10%
    2) 20%
    3) 30%
    4) 40%
    5) 50%

25. **What is the ratio of the number of students taking Spanish to the total student body?**

    1) 1:2
    2) 3:10
    3) 2:5
    4) 2:1
    5) 5:2

26. **What is the ratio of the number of students taking French to the number of students not taking French?**

    1) 3:10
    2) 3:7
    3) 3:5
    4) 5:3
    5) 7:3

27. In the figure below, the ratio of the vertical side to the horizontal side is the same in both triangles. If the height of the large triangle is 9" and its base is 12", what is the height of the small triangle if its base is 4"?

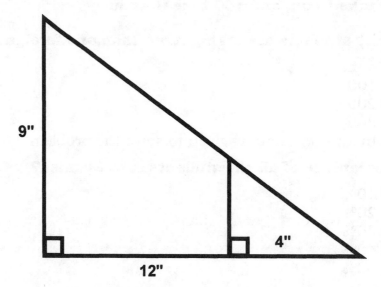

9"

12"

4"

   1) 1"
   2) 2"
   3) 3"
   4) 4"
   5) 5"

28. A container holds 7 $\frac{1}{2}$ gallons of water. What is the weight of the water in the container if 1 gallon of water weighs 8 $\frac{1}{3}$ pounds?

   1) 56 $\frac{1}{5}$ lbs.

   2) 56 $\frac{1}{6}$ lbs.

   3) 56 $\frac{5}{6}$ lbs.

   4) 62 $\frac{1}{6}$ lbs.

   5) 62 $\frac{1}{2}$ lbs.

170

29. Scarlette Knight starts out her act with 24 balloons. Half of the balloons break almost immediately, and by the end of the act a third of those remaining break as well. In total, what percentage of the balloons made it through the entire act?

1) $16 \frac{2}{3}\%$

2) $33 \frac{1}{3}\%$

3)   50%

4) $66 \frac{22}{3}\%$

5) $83 \frac{1}{3}\%$

30. John takes his new car out for a test spin and averages $27 \frac{1}{2}$ miles to the gallon. How many gallons did he use if he drove for a total of 143 miles?

1) 5.1 gallons
2) 5.2 gallons
3) 5.4 gallons
4) 52.0 gallons
5) 115.5 gallons

31. Complete the ad to the right with the correct percentage.

1) 12.5
2) 33.3
3) 62.5
4) 66.6
5) 75.0

____% off.

Find the lowest prices of the season at
Lingers Famous Semiannual Sale.
Beginning tomorrow.

Join us at our Semiannual Sale and discover
an additional 25% off the marked price of
50% off.* That's ____% off the original price
of our fall and winter classics. Shop now to take
advantage of the best prices on our quality clothing
(including petite sizes, shoes and accessories.
*Limited catalog styles may be marked at 1 /3 to 1 /2 off.

Gingers

The graph below to the right reflects housing starts from April 1986 through April 1987. Questions 32 and 33 refer to this graph.

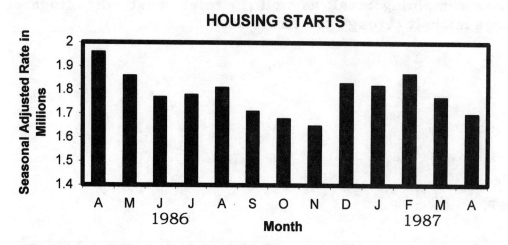

HOUSING STARTS

Source: U.S. Dept. of Commerce

32. What month had the fewest housing starts?

    1) June 1986
    2) September 1986
    3) November 1986
    4) January 1987
    5) April 1987

33. How many months had fewer than 1.75 million housing units started?

    1) 2
    2) 4
    3) 6
    4) 8
    5) 10

Questions 34 and 35 refer to the following situation:

The fire department is running a 50-50 raffle. Tickets sell for 50 cents each. 750 tickets are sold, and the prize is 50% of the proceeds.

34. How much will the winner receive?

    1) $ 93.75
    2) $187.50
    3) $375.00
    4) $562.50
    5) $750.00

35. If Bob spent $5.00 on tickets, what are the chances that he will win the raffle?

  1) 5 out of 750, $\dfrac{5}{750}$

  2) 10 out of 750, $\dfrac{10}{750}$

  3) 5 out of 745, $\dfrac{5}{745}$

  4) 10 out of 740, $\dfrac{10}{740}$

  5) 10 out of 375, $\dfrac{10}{375}$

36. Ilene has taken a part time job to save for a down payment on a new car. If the job pays $150.00 per week after deductions, and she wants to make a down payment of $3,450, how long will she have to work the second job?

  1) 13 weeks
  2) 20 weeks
  3) 21 weeks
  4) 23 weeks
  5) 24 weeks

Problems 37 through 40 refer to the following situation:

Dick fills his car with gas and notes that his odometer reads 21,071.6 miles. It takes 8 gallons to fill the tank, and it costs Dick $11.67. The next time he fills up, the price of gas has gone down 1 cent per gallon. The odometer reads 21,347 and it takes 10.2 gallons.

37. What is the price per gallon at the first stop?

  1) $1.13 $\dfrac{4}{10}$

  2) $1.14 $\dfrac{4}{10}$

  3) $1.14 $\dfrac{9}{10}$

  4) $1.44 $\dfrac{9}{10}$

  5) $1.45 $\dfrac{9}{10}$

173

38. How far did Dick travel between stops?

       1) 274.6 miles
       2) 275.4 miles
       3) 275.6 miles
       4) 276.4 miles
       5) 276.6 miles

39. How much does it cost to fill up the tank at the second stop?

       1) $11.59
       2) $11.88
       3) $14.49
       4) $14.78
       5) $26.37

40. How many miles per gallon did Dick get between visits?

       1) 2.69
       2) 2.70
       3) 26.90
       4) 27.00
       5) 27.10

41. Bob (you don't have to call me "Flash") Floyd finishes a 10K (kilometer) race in 46 minutes and 30 seconds. If one kilometer equals .62 miles, what was his time per mile?

       1) 7 min. 30 sec.
       2) 7 min. 47 sec.
       3) 7 min. 50 sec.
       4) 13 min. 20 sec.
       5) 13 min. 30 sec.

Questions 42 through 45 refer to the following graph.

Monthly Take-Home Pay = $800

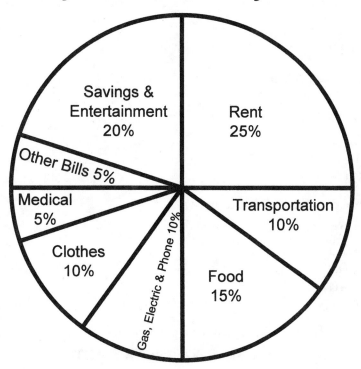

42. How much is spent on rent each month?

    1) $25
    2) $250
    3) $100
    4) $200
    5) $400

43. What item in the budget is more expensive then the rent?

    1) food
    2) clothes
    3) gas & electric
    4) medical & transportation
    5) none of the above

44. How much more is spent on food than on transportation?

    1) $120
    2) $80
    3) $40
    4) $5
    5) $200

45. If $60 is saved in one month, how much is spent on entertainment?

    1) $100
    2) $160
    3) $80
    4) $150
    5) $140

46. A 15% discount on a $60 item would be

    1) $9
    2) $3
    3) $6
    4) $12
    5) $20

47. Find a 20% tip on a $47 dinner.

    1) $2.35
    2) $4.70
    3) $7.05
    4) $9.40
    5) $11.75

Problems 48 - 51 refer to the following graph.

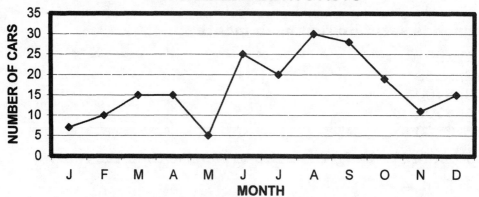

48. During which period of time is there the least change?

    1) January-February
    2) March-April
    3) May-June
    4) July-August
    5) September-October

49. **Peak sales were reached In which month?**

    1) February
    2) June
    3) July
    4) August
    5) December

50. **How many more sales were there in June than in December?**

    1) 10
    2) 15
    3) 5
    4) 20
    5) 25

51. **The longest period without a decline in sales was**

    1) 1 month
    2) 2 months
    3) 3 months
    4) 4 months
    5) 5 months

52. **28 of 35 students passed a test. What percent failed?**

    1) 7%
    2) 28%
    3) 20%
    4) 25%
    5) 80%

53. **A shirt which was discounted 15% is on sale for $17.00. The original price was**

    1) $20.00
    2) $19.55
    3) $14.00
    4) $14.55
    5) $15.00

54. **If there are 3 math teachers for every 85 math students in a particular school district, how many total math students are there if the district employs 12 math teachers?**

    1) 340
    2) 85
    3) 255
    4) 850
    5) 170

55. If a student can read $\frac{1}{6}$ of a book in $1\frac{1}{2}$ hours, how long would it take her to read half the book?

      1) 3 hrs.

      2) $3\frac{1}{2}$ hrs.

      3) 4 hrs.

      4) $4\frac{1}{2}$ hrs.

      5) 5 hrs.

56. After 10 weeks of the season, 22 of the 28 teams in the league are still alive for a playoff spot. What percentage does this represent?

      1) 60.4%
      2) 78.6%
      3) 72.8%
      4) 81.8%
      5) 70.5%

Word Problem Solutions

1. 2) Lauren receives a 25% raise or $\frac{1}{4}$ of $1,600. Dividing 4 into 1,600 is the easiest way to find the raise of $400. Lauren's gross income is $2,000. The company keeps a quarter of this. This time we divide 4 into 2,000 to find that withholding is $500. That leaves Lauren with a net income of $1,500.

   The correct answer is 2).

2. 5) If she shares an apartment, she will pay 20% of her gross pay of $2,000. 10% of $2,000 is $200. 20% would be twice this amount of $400.

   The correct answer is 5).

3. 3) If she lives on her own, it will cost Lauren $\frac{1}{3}$ of her take home pay of $1,500. By dividing 3 into $1,500, we find that her rent would be $500.

   The correct answer is 3).

4. 1) The savings on rent would be the difference between $500 and $400.

   The correct answer is 1).

5. 4) $10 each week for gas for 50 weeks comes to $500. $1.00 a day for tolls comes to $5.00 per week and a total of $250 for the year. Finally, there is the additional $150 Lauren must spend on car insurance. This comes to a total of $900.

   The correct answer is 4).

6. 5) $400 each month on rent comes to $4,800 in a year. Add to this the $900 calculated in problem 5, and the total expense is $5,700.

   The correct answer is 5).

7. 4) At $500 per month, the single apartment would cost $6,000 per year. This is $300 more than the shared apartment.

   The correct answer is 4).

8. 4) Each window measures 1 yard square. Converting this to feet and finding the area, we would get 3' x 3' = 9 sq. ft. The area of the closet door is 2' x 7' = 14 sq. ft. The area of the entrance door is 3' x 7' = 21 sq. ft. The total surface area of all doors and windows is the sum of 9, 9, 14, and 21, or 53 sq. ft.

   The correct answer is 4).

9. 3) They have to paint 4 walls. Two are 12 feet long and 8 feet high. Each is 96 sq. ft. The other two walls are 10 feet long and 8 feet high. Each is 80 sq. ft. The total area of the walls is:

2 x 96 plus 2 x 80 = 352 sq. ft.

The area that they have to paint is the difference between the total area of the walls (352) and the area taken up by doors and windows (53). 352 − 53 = 299 sq. ft.

The correct answer is 299 sq. ft. or 3).

10. 3) One gallon will be enough. They will pay the sale price which is 25% off of $16.40. Dividing 4 into 16.40, we find the discount of $4.10. Be careful, though, this is only the discount. The problem asks for the sale price, and that's $12.30.

The correct answer is 3).

11. 1) Now the correct answer is the discount of $4.10.

The correct response is 1).

12. 2) We are interested in the perimeter of the property which is 33 yds. long and 50 yds. deep. The perimeter is 166 yards. The cost of the fence is 166 times $2.10 or $348.60.

The correct answer is 2).

13. 3) 18 cans would cost $6.00. The additional 2 cans would cost $\frac{2}{3}$ of $1.00 or 67 cents. The total tab is $6.67.

The correct answer is 3).

14. 2) The first week Tommy earns $30, the second $100, the third nothing, and the fourth week $50. He earns a total of $180 for the four weeks. By dividing 180 by 4, we find that the average is $45 per week.

The correct answer is 2).

15. 3) Tommy works two hours a day, 5 days a week at the parts store for a total of 10 hours per week. 10 times $8 comes to $80 each week or $320 for the 4 weeks. In addition, he worked one Saturday for 8 hours and earned an extra $64. His total for the month was $384.

The correct answer is 3).

16. 3) Tommy's total monthly income from both sources is $180 plus $384 or $564. Dividing by 4, we find the average is $141. Another way of finding the answer would have been to divide 384 by 4 to get the average earnings at the auto parts store. Taking the result (96) and adding it to the tutoring average (45) you would again get $141.

The correct answer is 3).

17. 5)   Week #1: $80 at the store and $30 tutoring for a total of $110.
Week #2: $80 at the store and $100 tutoring for a total of $180.
Week #3: $144 at the store and nothing tutoring.
Week #4: $80 at the store and $50 tutoring for a total of $130.
The correct order is 2, 3, 4, 1.

The correct response is 5).

18. 2)   The list price is 138.9 cents per gallon. Multiplying this by 10.5 we get:

10.5 x 138.9 = $14.58

She gets a 2% discount.

98% * $14.58 = .98 * $14.58 = $14.29

or

2% * $14.58 = .02 * $14.58 = $0.29

$14.58 - $.29 = $14.29

The correct answer is 2).

19. 3)   Lois saves two percent. On a purchase of $14.58.

2% * $14.58 = .02 * $14.58 = $0.29

The correct answer is 3).

20. 3)   The entire perimeter is 2 times 11' 4" (22' 8") plus 2 times 8.5' (17') or 39' 8". Jack is about to answer 4), but at the last minute he realizes that he doesn't need molding for the 3 foot entrance. He only needs 36' 8" or $36 \frac{2}{3}$ ft.

The correct answer is 3).

21. 4)   If 100 eggs cost $9.00, then the price per egg is 9 cents. Twelve eggs, therefore, cost $1.08.

The correct answer is 4).

22. 3)   The total enrollment in foreign language courses is equal to the sum of 400, 300, 200, and 100 or 1000. If only 800 individuals are taking a foreign language and no one takes more than two, then 1000 minus 800 or 200 students are taking two languages.

The correct answer is 3).

23. 4) $\frac{400}{1000}$ reduces to $\frac{40}{100}$ which is 40%.

The correct answer is 4).

24. 5)   The key here that makes it different from the preceding problem is that we are only interested in the percentage of 800 students taking a language. The fraction is $\frac{400}{800}$ or 50%.

The correct answer is 5).

25. 3)   The ratio is 400 to 1000. Dividing both numbers by 200, the ratio reduces to 2 to 5.

The correct answer is 3).

26. 2)   Out of 1000 students, only 300 take French. That means 700 don't take French. The ratio is 300 to 700. This reduces to 3 to 7.

The correct answer is 2).

27. 3) $$\frac{9}{12} = \frac{?}{4}$$

12 x ? = 4 x 9
12 x ? = 36

$$? = \frac{1}{12} \text{ x } 36$$

? = 3

The correct answer is 3).

28. 5)   The product of $7\frac{1}{2}$ and $8\frac{1}{3}$ is:

$$\frac{15}{2} \text{ x } \frac{25}{3} = \frac{375}{6} = \frac{125}{2} = 62\frac{1}{2}$$

You could have saved one step in the problem if you canceled before performing the multiplication. 3 divides evenly into 15 and 3.

The correct answer is 5).

29. 2)   Initially, Scarlette loses half or 12 of the 24 balloons. Of the remaining 12, she loses a third or 4 of these. That leaves her with 8 balloons out of the original 24. $\frac{8}{24}$ equals $\frac{1}{3}$ and that's $33\frac{1}{3}$%.

The correct answer is 2).

30. 2)   Dividing 27.5 into 143 we get:

$$27.5\overline{)143} = 275\overline{)1430.0}^{\,5.2}$$

The correct answer is 2).

31. 3)   To make it easy, let's assume that the original price of an item to be purchased is \$100. 50% off would be \$50. 25% off the remaining \$50 would amount to an additional \$12.50 savings ($\frac{1}{4}$ of 50 dollars). The total discount is \$62.50 off of the original \$100. $\frac{62.5}{100}$ = 62.5%.

The correct answer is 3).

32. 3)   The shortest bar is N 1986.

The correct answer is November 3).

33. 2)   If you draw a horizontal line through the graph at 1.75, it is obvious that 3 and maybe even 4 bars fall below the line. It is, in fact, 4, but even if you're personal choice was between 3 and 4, notice that the test response restricts you to 2 or 4. You can therefore assume that April of 1987 does fall below the line and that the correct response is 2.

The correct answer is 2).

34. 2)   750 tickets are sold at 50 cents each, so the total amount taken in is \$375. The winner will get half of this, or \$187.50.

The correct answer is 2).

35. 2)   If Bob spent \$5.00 on tickets, he purchased ten of the 750 tickets sold. He therefore has 10 out of 750 chances of winning.

The correct answer is 2).

36. 4)   Dividing 150 into 3,450 we get:

$$150\overline{)3450} \quad 23$$

The correct answer is 4).

37. 5)   Dick gets 8 gallons and it costs him \$11.67. Dividing 8 into \$11.67 we get:

$$8\overline{)11.67000} = 145\frac{9}{10} \quad 1.45875$$

The correct answer is 5).

38. 2)

$$\begin{array}{r} 21347.0 \\ \underline{21071.6} \\ 275.4 \end{array}$$

The correct answer is 2).

39. 1)   Gas cost $1.45 $\frac{9}{10}$ per gallon at the first stop, so the second stop cost

$1.44 $\frac{9}{10}$ per gallon. 10.2 gallons would cost:

1.449 x 10.2 = 14.78

The correct answer is 4).

40. 4)   He went 275.4 miles on 10.2 gallons of gas. He averaged:

$$10.2\overline{)275.4} = 102\overline{)2754}^{\,27}$$

The correct answer is 4).

41. 1)   Bob ran 6.2 miles in 46.5 minutes. Dividing 6.2 into 46.5 we get:

$$6.2\overline{)46.5} = 62\overline{)465.0}^{\,7.5}$$

7.5 minutes equals 7 minutes and 30 seconds.

The correct answer is 1).

42. 4)   25% of $800 = 800 x .25 = $200.

The correct answer is 4).

43. 5)   Rent is largest item at 25%.

The correct answer is 5).

44. 3)   Food is 15% and transportation is 10%. The difference is 5%. 5% of $800 is $40.

The correct answer is 3).

45. 1)   Savings and entertainment represent 20% of the budget. 20% of $800 is $160. If $60 is saved, then $160 - $60 or $100 goes towards entertainment.

The correct answer is 1).

46. 1)   10% of $60 is $6.00. 5% would be half of this or $3.00. 15% = 10% + 5% = $9.00.

The correct answer is 1).

47. 4)   10% of $47 is $4.70. Twice that would be $9.40.

The correct answer is 4).

48. 2)   A horizontal line runs from March to April. This is the period of least change.

The correct answer is 2).

49. 4) The peak of the graph, 30 sales, occurs in August.

The correct answer is 4).

50. 1) June sales were 25 and December sales were 15. The difference is 10.

The correct answer is 1).

51. 3) Sales did not decline from January to April – a period of 3 months.

The correct answer is 3).

52. 3) If 28 passed, then 7 failed.

$$\frac{7}{35} = \frac{1}{5} = 20\%$$

The correct answer is 3).

53. 1) If the discount was 15%, then $17.00 represents 85% of the original cost.

$$17 \div 85\% = .85\overline{)17.00} = 85\overset{20}{\overline{)1700}} = 20$$

The correct answer is 1).

54. 1)

$$\frac{3}{85} = \frac{12}{?} = \frac{12}{340}$$ 3 goes into 12, 4 times

4 times 85 = 340

The correct answer is 1).

55. 4) $\frac{1}{2}$ is equal to $\frac{3}{6}$, so it would take 3 times as long to read.

$$3 \times 1\frac{1}{2} = 3 \times \frac{3}{2} = \frac{9}{2} = 4\frac{1}{2} \text{ hrs.}$$

The correct answer is 4).

56. 2) $\dfrac{22}{28} = \dfrac{22 \div 2}{28 \div 2} = \dfrac{11}{14}$

$$14\overset{.7857}{\overline{)11.0000}} = 78.57\% \text{ or approximately } 78.6\%$$

The correct answer is 2).

NOTES

186

Mathematics Test Answer Sheet Part I

1. ① ② ③ ④ ⑤
2. ① ② ③ ④ ⑤
3. ① ② ③ ④ ⑤
4. ① ② ③ ④ ⑤
5. ① ② ③ ④ ⑤
6. ① ② ③ ④ ⑤
7. ① ② ③ ④ ⑤
8. ① ② ③ ④ ⑤
9. ① ② ③ ④ ⑤
10. ① ② ③ ④ ⑤
11. ① ② ③ ④ ⑤
12. ① ② ③ ④ ⑤

15. ① ② ③ ④ ⑤
16. ① ② ③ ④ ⑤
17. ① ② ③ ④ ⑤
18. ① ② ③ ④ ⑤
19. ① ② ③ ④ ⑤

13.

	/	/	/	
•	•	•	•	•
0	0	0	0	0
1	1	1	1	1
2	2	2	2	2
3	3	3	3	3
4	4	4	4	4
5	5	5	5	5
6	6	6	6	6
7	7	7	7	7
8	8	8	8	8
9	9	9	9	9

20.

	/	/	/	
•	•	•	•	•
0	0	0	0	0
1	1	1	1	1
2	2	2	2	2
3	3	3	3	3
4	4	4	4	4
5	5	5	5	5
6	6	6	6	6
7	7	7	7	7
8	8	8	8	8
9	9	9	9	9

21.

	/	/	/	
•	•	•	•	•
0	0	0	0	0
1	1	1	1	1
2	2	2	2	2
3	3	3	3	3
4	4	4	4	4
5	5	5	5	5
6	6	6	6	6
7	7	7	7	7
8	8	8	8	8
9	9	9	9	9

22.

	/	/	/	
•	•	•	•	•
0	0	0	0	0
1	1	1	1	1
2	2	2	2	2
3	3	3	3	3
4	4	4	4	4
5	5	5	5	5
6	6	6	6	6
7	7	7	7	7
8	8	8	8	8
9	9	9	9	9

23. ① ② ③ ④ ⑤
24. ① ② ③ ④ ⑤
25. ① ② ③ ④ ⑤

14.

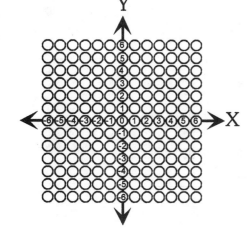

Mathematics Test Answer Sheet Part II

26. ① ② ③ ④ ⑤
27. ① ② ③ ④ ⑤

28.

29.

30.

31.

32.

33. ① ② ③ ④ ⑤
34. ① ② ③ ④ ⑤
35. ① ② ③ ④ ⑤
36. ① ② ③ ④ ⑤
37. ① ② ③ ④ ⑤
38. ① ② ③ ④ ⑤
39. ① ② ③ ④ ⑤
40. ① ② ③ ④ ⑤
41. ① ② ③ ④ ⑤
42. ① ② ③ ④ ⑤
43. ① ② ③ ④ ⑤
44. ① ② ③ ④ ⑤
45. ① ② ③ ④ ⑤
46. ① ② ③ ④ ⑤
47. ① ② ③ ④ ⑤
48. ① ② ③ ④ ⑤
49. ① ② ③ ④ ⑤
50. ① ② ③ ④ ⑤

FORMULAS

AREA of a:

Square	Area = side^2
Rectangle	Area = length • width
Parallelogram	Area = base • height
Triangle	Area = ½ • base • height
Trapezoid	Area = ½ • base$_1$ • base$_2$ • height
Circle	Area = π • radius^2, π is approximately equal to 3.14

PERIMETER of a:

Square	Perimeter = 4 • side
Rectangle	Perimeter = 2 • length + 2 • width
Parallelogram	Perimeter = 2 • side$_1$ + 2 • side$_2$
Triangle	Perimeter = side$_1$ + side$_2$ + side$_3$
Trapezoid	Perimeter = side$_1$ + side$_2$ + side$_3$ + side$_4$
Circle	Circumference = π • diameter, π is approximately equal to 3.14

VOLUME of a:

Cube	Volume = edge^3
Rectanglar Solid	Volume = length • width • height
Pyramid	Volume = $\frac{1}{3}$ base • height
Cylinder	Volume = π • radius^2 • height
Cone	Volume = $\frac{1}{3}$ π • radius^2 • height, π is approximately equal to 3.14

COORDINATE GEOMETRY

distance between points (x_1, y_1) and (x_2, y_2)

$$\sqrt{(x_2 - x_1)^2 + (y_2 - y_1)^2}$$

Slope of a line = $\frac{(y_2 - y_1)}{(x_2 - x_1)}$ where (x_1, y_1) and (x_2, y_2) are two points on a line

PYTHAGOREAN THEOREM

$a^2 + b^2 = c^2$; a and b are the sides and c is the hypotenuse of a right triangle.

AVERAGE

mean = $\frac{x_1 + x_2 + x_3 \ldots + x_n}{n}$ where the x's are the values and n is the total number of values.

median = the middle value in a series

SIMPLE INTEREST

interest = principal x rate x time

DISTANCE

distance = rate x time

TOTAL COST

total cost = (price per unit) x (number of units)

Calculator Directions

Turn the calculator on by pressing the $\boxed{\text{ON}}$ key. DEG appears in small letters in the top center of the screen. 0. Appears at the right. This is means the calculator is in the proper format for all your calculations.

To clear information from previous calculations when working on questions hit the $\boxed{\text{AC}}$ key.

To do arithmetic operations, enter them as written. Press $\boxed{=}$ when finished.

EXAMPLE: 3 + 4 − 1 =

>   Clear the calculator by pressing $\boxed{\text{AC}}$
>
>   Then press $\boxed{3}$ $\boxed{+}$ $\boxed{4}$ $\boxed{-}$ $\boxed{1}$ $\boxed{=}$
>
>   The correct answer is 6

To multiply a number by an expression in parentheses, press x between the number and the parentheses sign.

EXAMPLE: 3(4 - 2) =

>   Clear the calculator by pressing $\boxed{\text{AC}}$
>
>   Then press $\boxed{3}$ $\boxed{\text{x}}$ $\boxed{(}$ $\boxed{4}$ $\boxed{-}$ $\boxed{2}$ $\boxed{)}$ $\boxed{=}$
>
>   The correct answer is 6

To find the square root of a number, enter the number; press $\boxed{\text{shift}}$ shift appears at the top left of the screen; press $\boxed{x^2}$ (top row 3rd from left) this accesses the square root function.

EXAMPLE: $\sqrt{81}$

>   Clear the calculator by pressing $\boxed{\text{AC}}$
>
>   Then press $\boxed{8}$ $\boxed{1}$ $\boxed{\text{shift}}$ $\boxed{x^2}$ $\boxed{=}$
>
>   The correct answer is 9

To enter a negative number, enter the number; press the change sign key $\boxed{+/-}$ which is above the $\boxed{7}$ key.

EXAMPLE: -5 + 14

>   Clear the calculator by pressing $\boxed{\text{AC}}$
>
>   Then press $\boxed{5}$ $\boxed{+/-}$ $\boxed{+}$ $\boxed{14}$ $\boxed{=}$
>
>   The correct answer is 9

You may use a calculator for this part of the test.

1. Colleen is a Daisy leader and she budgets $15 as initial startup cost for art supplies. At the supply store, she sees that colored paper sells for $1.25 per pack, a package of 5 small scissors goes for $3.75 per pack, and glue sticks go for $.75 each. If she buys 2 packs of colored paper, 10 small pairs of scissors, how many glue sticks can she buy?

   (1) 4
   (2) 5
   (3) 6
   (4) 7
   (5) 8

2. Michelle earns $7.50/hour babysitting. Last month she worked for a total of $13 \frac{1}{2}$ hours. With the money she earned, she goes clothes shopping. She buys a skirt for $17.30, a turtleneck sweater for $19.75, a pair of shoes for $24.95, a hat for $23.50, and two pairs of stockings for $5.95 each. How much money does she have left from her babysitting?

   (1) $1.35
   (2) $3.85
   (3) $9.80
   (4) $12.00
   (5) $15.75

3. Allison, Sandi, and Mandy decide to get in some softball practice in Allison's backyard. They spread out and form the triangle below. How far does Mandy have to throw to reach Allison?

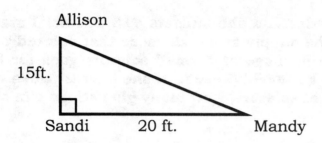

    (1)   15 ft.
    (2)   20 ft.
    (3)   25 ft.
    (4)   30 ft.
    (5)   35 ft.

4. Hannah is in charge of buying soda for the school dance. They sold 672 tickets and she figures that each person will drink 3 cans of soda. If there are 24 cans to a case, how many cases of soda should Hannah buy?

    (1)   10
    (2)   28
    (3)   56
    (4)   84
    (5)   224

5. If the distance between two points is represented by the formula $\sqrt{(X_2 - X_1)^2 + (Y_2 - Y_1)^2}$, what is the length of the line segment connecting the two points below?

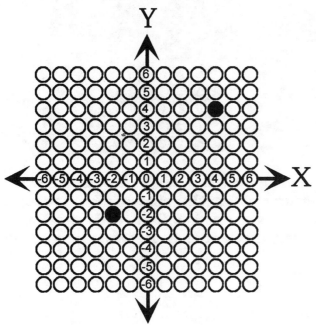

(1) 2,83
(2) 5.66
(3) 8.49
(4) −5.66
(5) −2.83

6. Romeo plans to elope with Juliet. Her bedroom window is 14 feet from the ground. Running along the side of the house, underneath the window is her father's prized 8-foot wide flower garden and Romeo must avoid putting the ladder in it. How long must his ladder be?

        (1)   14'
        (2)   22'
        (3)   15.53'
        (4)   16.12'
        (5)   16'

Questions 7, 8 and 9 refer to the following information

Marissa goes to the toy store and has $1.00 to spend on marbles. The store sells two different bags of marbles, but the customers are also allowed to mix and match and make up their own bag.

	Fraction of Blue Marbles	Number of Marbles in Bag	Cost
Bag 1	$\frac{3}{8}$	24	All marbles cost 10 cents
Bag 2	$\frac{1}{5}$	25	20 Cents for Blue Marbles 10 Cents for Yellow Marbles 15 Cents for Red Marbles

7. **If Marissa were to buy all of the blue marbles in each of the bags, how much money would she need?**

      (1) $.90
      (2) $1.00
      (3) $1.40
      (4) $1.90
      (5) Not enough information is given

8. **How much money would Marissa need to buy both bags of marbles?**

      (1) $4.50
      (2) $5.40
      (3) $1.90
      (4) $4.90
      (5) Not enough information is given.

9. **Marissa buys as many of the blue marbles from the first bag as she can afford. With her change, she buys yellow marbles from the second bag. Assuming there are enough yellow marbles to select from, how many can she buy?**

      (1) 0
      (2) 1
      (3) 2
      (4) 3
      (5) Not enough information is given.

10. In the town softball league, teams are made up of fifth and sixth graders. For each team, there are 3 fifth-graders in each starting lineup of 10 girls. On any given Saturday, there are 6 games scheduled. In total, how many fifth-graders are in the starting lineups?

        (1) 3
        (2) 6
        (3) 18
        (4) 24
        (5) 36

Questions 11 and 13 refer to the following data.

Brady has set up a jogging route in his neighborhood. The map below represents the blocks that he runs on and the distances that he has been able to measure. Cedar Ridge Road crosses Clover Hill Road and connects Cider Mill Road and Morgan Road, forming two similar triangles. Brady is able to measure the distance of Morgan Rd (.3 miles), Cider Mill Rd (.45 miles), and he knows that the total length of Clover Hill is 1.5 miles.

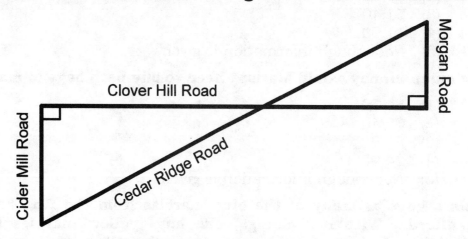

11. **What other intersection forms the same angle as the intersection of Cider Mill and Cedar Ridge?**

        (1)   Cedar Ridge and Clover Hill
        (2)   Cider Mill and Clover Hill
        (3)   Clover Hill and Morgan
        (4)   Cedar Ridge and Morgan
        (5)   There are no other intersections with the same angle.

12. **What is the shortest distance from Cider Mill to where Cedar Ridge intersects Clover Hill ?**

       (1)   .375 miles
       (2)   .5 miles
       (3)   .6 miles
       (4)   .9 miles
       (5)   1.2 miles

13. **How long is Cedar Ridge Road? Round your answer to the nearest thousandth.**

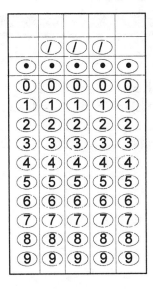

14. **On the grid below darken the midpoint of a line that connects the points (-1 , 2) and (5 , 6).**

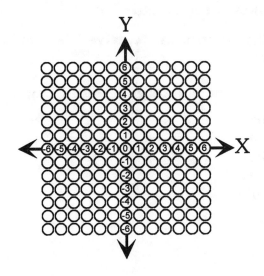

197

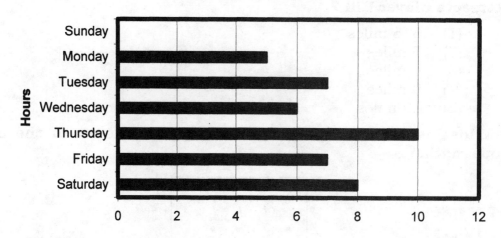

Tina's Hours Last Week

15. **Tina earns $10.75 an hour for normal pay. For overtime hours (hours over 40 total in a week) she earns time and a half or 50% more per hour. If 18% of her pay is withheld to pay for various taxes, how much can she expect to be in her check when she picks it up this Friday?**

    (1) $86.11
    (2) $360.56
    (3) $392.27
    (4) $462.25
    (5) $478.38

Questions 16 through 18 refer to the following data.

George commutes each day from Flemington to Jersey City, a distance that measures $6\frac{2}{3}$ inches on a map that is drawn to a scale of 1 inch = 7.5 miles. On Saturday George must drive his mother to visit his sister a round trip of 240 miles.

16. **Approximately, how far is George's round trip commute each day?**

    (1)   42 miles
    (2)   50 miles
    (3)   84 miles
    (4)   93 miles
    (5)   100 miles

17. **If George gets 20 miles/gallon in his car, approximately how many gallons of gas will he use driving his mother Saturday?**

    (1)   2 gallons
    (2)   5 gallons
    (3)   6 gallons
    (4)   12 gallons
    (5)   24 gallons

18. **George's mother insists on paying him 5 cents a mile. He accepts the money to avoid the fight. If gas costs $1.47/gallon, approximately how much will George's mother pay him?**

    (1) $6.47
    (2) $7.35
    (3) $12.00
    (4) $17.64
    (5) $88.20

19. **What was the original price of a watch that sold at a 25% discount for $69.00?**

    (1) $92.00
    (2) $17.25
    (3) $86.25
    (4) $115.00
    (5) $23.00

Questions 20, 21, and 22 refer to the following information. Use your calculator and use the grids to show your answers.

Sandi walks along a riverbank from point A to point B where she spots an apple tree on the other side of the river at point C. She can either wade directly across the river the river to point C, or she can continue to walk to point D where there is a bridge that spans the river to point E. We know the following distances:

The distance from A to B is 2 kilometers, from B to C is 2.5 kilometers, and point B is halfway between A and D.

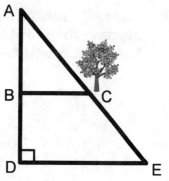

20. How many kilometers is it from point A to point C? Express your answer to the nearest tenth of a kilometer.

21. How many times farther is it from point D to point E than from point B to point C?

22. If 1 kilometer equals .62 miles, how many miles will Sandi travel if she walks from A to D, crosses the river at the bridge, and then returns to A from E? Express your answer to the nearest hundredth of a mile

Questions 23, 24, and 25 refer to the following information

In order to convert a temperature reading in Fahrenheit to one measured in Celsius, subtract 32 from the Fahrenheit measurement and multiply the result by $\frac{5}{9}$.

23. **If F represents the measurement in Fahrenheit and C represents the measurement in Celsius, which of the following correctly represents the relationship expressed above?**

    (1) $C = \frac{5}{9}(F - 32)$

    (2) $C = \frac{5}{9}F - 32$

    (3) $C = \frac{5}{9}(F + 32)$

    (4) $C = \frac{5}{9}F + 32$

    (5) $C = \frac{9}{5}F - 32$

24. **What is the Celsius temperature reading that corresponds to 131 degrees Fahrenheit?**

    (1) 40°
    (2) 55°
    (3) 72.5°
    (4) 90°
    (5) 99°

25. **What would be the correct formula for representing Fahrenheit as a function of Celsius?**

    (1) $F = \frac{9}{5}(C - 32)$

    (2) $F = \frac{9}{5}C - 32$

    (3) $F = \frac{9}{5}(C + 32)$

    (4) $F = \frac{5}{9}(C + 32)$

    (5) $F = \frac{9}{5}C + 32$

NOTES

NOTES

You may **NOT** use a calculator for this part of the test.

Questions 26 and 27 refer to the following information:

Bob and Mike decide to try their hand at indoor painting as a summer job. The first day on the job, in 7 $\frac{1}{2}$ hours they were able paint a total of 3 rooms.

26. **If Bob paints twice as many walls as Mike, and assuming each room had 4 walls, how many walls did Mike paint?**

    (1)   4
    (2)   6
    (3)   8
    (4)   10
    (5)   12

27. **If the boys start working 10 hours per day and keep up the same pace, how many rooms could they paint in a 5-day work week?**

    (1)   15
    (2)   20
    (3)   21
    (4)   28
    (5)   60

28. **On the grid below, darken in the circle that represents the point (-2,5).**

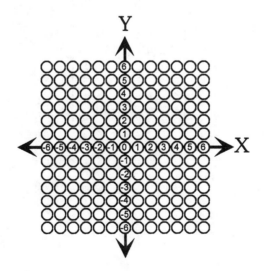

Questions 29 through 32 refer to the following data. Use the grids to show your answers.

Fine Grind Processed Hardwood Bark costs $16.00 per cubic yard and Cedar Mulch costs $25.00 per cubic yard and your order must be in whole cubic yards. Bob wants to replace the grass under the girls' swing set. He plans on outlining the area with railroad ties and fill in the 20' by 16' to a depth of 4 inches with landscaping mulch. In addition, Bob want to get 4 cubic yards of for his flower beds.

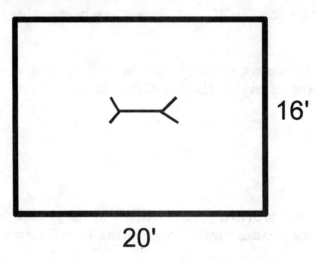

16'

20'

29. How many cubic yards of mulch must Bob buy for the swing set area?

30. How much more expensive would it be to use Cedar Mulch instead of Hardwood Bark on the flower beds?

	/	/	/	
⊙	⊙	⊙	⊙	⊙
⓪	⓪	⓪	⓪	⓪
①	①	①	①	①
②	②	②	②	②
③	③	③	③	③
④	④	④	④	④
⑤	⑤	⑤	⑤	⑤
⑥	⑥	⑥	⑥	⑥
⑦	⑦	⑦	⑦	⑦
⑧	⑧	⑧	⑧	⑧
⑨	⑨	⑨	⑨	⑨

31. If Bob decides on Cedar Mulch for the flower beds, how much will this cost?

	/	/	/	
⊙	⊙	⊙	⊙	⊙
⓪	⓪	⓪	⓪	⓪
①	①	①	①	①
②	②	②	②	②
③	③	③	③	③
④	④	④	④	④
⑤	⑤	⑤	⑤	⑤
⑥	⑥	⑥	⑥	⑥
⑦	⑦	⑦	⑦	⑦
⑧	⑧	⑧	⑧	⑧
⑨	⑨	⑨	⑨	⑨

32. If a wheelbarrow holds 3 cubic feet of mulch, how many trips will Bob make moving the mulch from his driveway his flower beds?

	/	/	/	
⊙	⊙	⊙	⊙	⊙
⓪	⓪	⓪	⓪	⓪
①	①	①	①	①
②	②	②	②	②
③	③	③	③	③
④	④	④	④	④
⑤	⑤	⑤	⑤	⑤
⑥	⑥	⑥	⑥	⑥
⑦	⑦	⑦	⑦	⑦
⑧	⑧	⑧	⑧	⑧
⑨	⑨	⑨	⑨	⑨

Questions 33, 34, and 35 refer to the following information

John is thinking of changing his phone service provider and has compiled the following information in the chart below:

	Company A	Company B	Company C	Company D	Company E
Per Minute Rates for State to State Calls:					
Off Peak	7¢	5¢	5¢	6¢	5¢
Peak	7¢	25¢	10¢	9¢	9¢
For In-State Calls:	7.9¢	11¢	9.8¢	5¢	9.8¢
Free Minutes Trial	0	1000	250	0	0

33. **John does not make any Out of State calls, so the company offering the best plan for his calling needs is**

    (1)   A
    (2)   B
    (3)   C
    (4)   D
    (5)   E

34. **John will use the cell phone exclusively for business to make Off Peak State to State calls and selects Company C over Company E because**

    (1)   The Off Peak rate is lower.
    (2)   He'll save a little money on the Peak calls he makes.
    (3)   They offered more free trial minutes.
    (4)   They provided web-based access to account information.
    (5)   He saves on Peak calls and they offered more free trial minutes.

35. John spends an equal amount of Peak and Off Peak time on the phone, so he is looking for the company that can provide the best average price for State to State calls. Since he also spends a few hours on the phone each month talking to his In-State friends, if he finds that there is more than one company that can provide the lowest average rate for out of state calls, he will select the one among them that has the lowest in-state rate. Which is the best plan for him?

    (1)   A
    (2)   B
    (3)   C
    (4)   D
    (5)   E

36. Ed digs a rectangular garden that is twice as long as it is wide. If he doubles the width and triples the length, by what factor will he have increased the area of the garden?

    (1)   Twice as large
    (2)   Three times as large
    (3)   Six times as large
    (4)   Ten times as large
    (5)   Twelve times as large

Questions 37, 38, and 39 refer to the graph below which represents the points scored per game by a professional football team during the 2000 regular season.

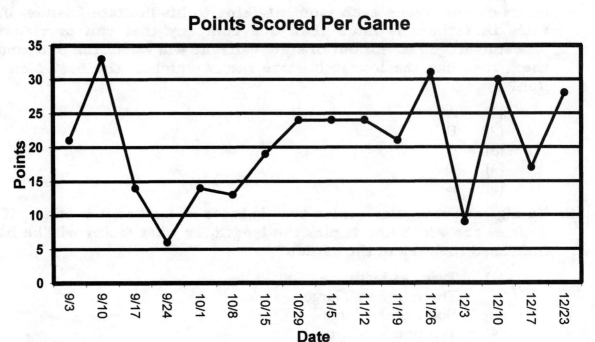

Points Scored Per Game

37. **What Point Value Represents the Mode?**

    (1)   20.5
    (2)   21
    (3)   24
    (4)   6
    (5)   33

38. **What was the most number of points scored in a game during the month of October?**

    (1)   14
    (2)   33
    (3)   31
    (4)   24
    (5)   21

39. **How many points did the team score in December?**

    (1)   115
    (2)   75
    (3)   328
    (4)   90
    (5)   84

Questions 40 and 41 refer to the following information

Michelle's dad wants to reward her whenever she brings home an excellent report card. At the beginning of the school year he gives her the option of either earning $4.00 for every A or $3.00 for the first 5 A's and $5.00 for each A after that for the rest of the school year. Michelle is graded in 5 subjects and gets four report cards each year, so she has the potential of 20 A's each school year.

40. If X equals the number of A's she receives and we know that she received at least 5 A's over the course of the year, then the two formulas could be represented as

      (1)   4X and 3X
      (2)   4X and 3X + 5(20 – X)
      (3)   4X and 3(5) + 5(X – 5)
      (4)   4X and 3X + 5(5 – X)
      (5)   4X and 3(X – 5) + 5X

41. How many A's must Michelle earn to receive the same amount of money under both calculation methods?

      (1)   5
      (2)   8
      (3)   10
      (4)   12
      (5)   15

Questions 42, 43, and 44 refer to the following information

One long distance phone service charges $1.00 for the first 20 minutes and $.07 for each minute after that. A second long distance phone service charges a flat 6 cents a minute rate.

42. If X represents the total length in minutes of a call that is longer than 20 minutes, which of the following formulas represents the total cost in dollars for a call placed under the first plan?

    (1)   X + 7X
    (2)   X + .07(X – 20)
    (3)   1 + .07(20 – X)
    (4)   1 + .07(X + 20)
    (5)   1 + .07(X - 20)

43. What's the break even point between the two plans?

    (1)   10 minutes
    (2)   20 minutes
    (3)   30 minutes
    (4)   40 minutes
    (5)   50 minutes

44. Under the first plan, how much would a person pay for a 47-minute call?

    (1) $1.75
    (2) $2.00
    (3) $2.35
    (4) $2.82
    (5) $2.89

Questions 45, 46 and 47 refer to the graph below which represents a consumers annual kilowatt usage by month as recorded by his electric company.

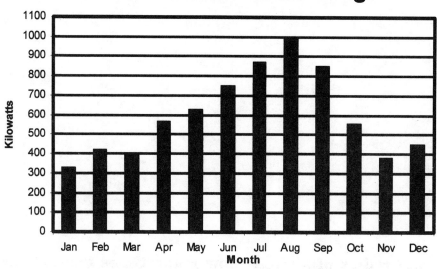

Annual Kilowatt Usage

45. **In which month did the consumer use the most electricity?**

    (1)   May
    (2)   June
    (3)   July
    (4)   August
    (5)   September

46. **In how many months was the usage less than February?**

    (1)   1
    (2)   2
    (3)   3
    (4)   4
    (5)   5

47. **What was the approximate total usage for the year?**

    (1)   6,600 kilowatts
    (2)   6,800 kilowatts
    (3)   7,000 kilowatts
    (4)   7,200 kilowatts
    (5)   7,400 kilowatts

48. The area of a circle is represented by the formula $A = \pi r^2$ and the Circumference is represented by the formula $C = \pi d$, where π is a constant, r is the radius of the circle and d is the diameter. Which of the following represents the ratio of the Area of a circle to the Circumference of a circle?

(1) $\dfrac{r}{2}$

(2) $\dfrac{2}{r}$

(3) $\dfrac{1}{1}$

(4) $\dfrac{r}{d}$

(5) $\dfrac{2r}{d}$

49. If you triple the radius of a circle, how many times greater does the area become?

(1) 3
(2) 6
(3) 9
(4) 12
(5) 15

50. If you double the radius of a circle, how many times greater does the circumference become?

(1) 2
(2) 3
(3) 4
(4) 8
(5) 16

NOTES

Practice Test – Part I Solutions

1. 3) Two packages of colored paper would cost 2 x $1.25 = $2.50 10 pairs of small scissors would be two packages. Each package costs $3.75, so the total cost of scissors would be 2 x $3.75 = $7.50

   Colleen's first two purchases total $10.00. This leaves her $5.00 ($15.00 - $10.00) to purchase glue sticks. Since each glue stick costs $.75, we have to divide .75 into 5.00. It goes 6 times with a remainder of 50 cents.

   The correct answer is 3).

2. 2) To find out how much money Michelle has to spend, we need to find out how much Michelle earned by multiplying her hourly rate ($7.50) by the number of hours she worked ($13 \frac{1}{2}$). You either have to change 7.5 to a fraction ($7 \frac{1}{2}$) or $13 \frac{1}{2}$ to a decimal (13.5). It's easiest to work with decimals, especially if you can use your calculator.

   $7.50 x 13.5 = $101.25   Michelle has $101.25 to spend.

   To find out how much money Michelle has spent, we have to add up all of her purchases.

1 skirt @ $17.30	=	$17.30
1 sweater @ $19.75	=	$19.75
1 shoes @ 24.95	=	$24.95
1 hat @ 23.50	=	$23.50
2 stockings @ 5.95	=	$11.90
		$97.40

   Michelle spent a total of $97.40 on her shopping spree.

   To find out how much money she has left, subtract the money spent ($97.40) from the total she began with ($101.25). The difference is $3.85.

   The correct answer is 2).

3. 3) The quickest way to come up with a solution to this problem is to recognize it as a 3-4-5 right triangle. Letting X = the distance between Allison and Mandy, we can set up the proportion:

$$\frac{3}{5} = \frac{15}{X}$$

$$\frac{3}{5} \nearrow\searrow \frac{15}{X}$$

3 goes into 15, 5 times. 5 times 5 equals 25. The distance is 25 feet.

The correct answer is 3).

4. 4) The total number of sodas that Hannah will need is the product of the number of people attending (672) and the number of sodas that each person will have (3).

672 x 3 = 2016 = total number of sodas

If there are 24 sodas in each case, then to determine the number of cases need, divide the total number of sodas (2016) by 24.

2016 ÷ 24 = 84.

The correct answer is 4).

5. 3) The two points are (-2, -2) and (4, 4). Applying the formula and using the square root function on our calculator, we get:

$$\sqrt{(X_2 - X_1)^2 + (Y_2 - Y_1)^2} = \sqrt{(4 - (-2))^2 + (4 - (-2))^2} = \sqrt{6^2 + 6^2} = \sqrt{72} = 8.49$$

The correct answer is 3).

6. 4) The figure is a right triangle with sides of 14' **and** 8'. Using the Pythagorean theorem we can solve for the length of the hypotenuse (the ladder). Using the X^2 and $\sqrt{\ }$ will make this easier as well as giving you practice in using the calculator.

The length of the ladder = $\sqrt{14^2 + 8^2}$ = $\sqrt{196 + 64}$ = $\sqrt{260}$ = 16.12

The correct answer is 4).

7. 4) The first bag contains 24 marbles of which $\frac{3}{8}$ are blue. To find the number of blue marbles in the first bag, you must find $\frac{3}{8}$ of 24. To do this, multiply 24 by $\frac{3}{8}$.

$$\frac{3}{8} \times \frac{24}{1} =$$

$$\frac{3}{1\cancel{8}} \times \frac{2\cancel{4}^{3}}{1} = 9$$

There are 9 blue marbles in the first bag. Each marble costs 10 cents. Therefore, the total cost of the 9 blue marbles in the first bag is equal to 9 x .10 = $.90.

The second bag contains 25 marbles of which $\frac{1}{5}$ are blue. To find the number of blue marbles in the second bag, you must find $\frac{1}{5}$ of 25. To do this, multiply 25 by $\frac{1}{5}$.

$$\frac{1}{5} \times \frac{25}{1} =$$

$$\frac{1}{1\cancel{5}} \times \frac{2\cancel{5}^{5}}{1} = 5$$

There are 5 blue marbles in the second bag. Each blue marble costs 20 cents. Therefore, the total cost of the 5 blue marbles in the first bag is equal to 5 x .20 = $1.00.

The total cost of all the blue marbles is equal to sum of these two costs. The sum of $.90 and $1.90 is $1.90.

The correct answer is 4).

8. 5) The first bag contains 24 marbles. Each marble costs 10 cents, regardless of color. The total cost of the bag would be the product of 24 (the total number of marbles) and .10 (the cost per marble).

24 x .10 = $2.40

To find the cost of the second bag, we must first find the cost for each color marble, and then sum each of these costs. In problem #1, we already found the cost of the blue marbles ($1.00). Now we need to find the cost of the red marbles and the yellow marbles. Since the red marbles are priced differently than the yellow marbles, we need to know exactly how many of each there are in the bag. However, unlike the blue marbles where we know they make up $\frac{1}{5}$ of the total count of 25, we don't know the fraction of yellow or red marbles. There is not enough information given to solve this problem.

The correct answer is 5).

9. 2) We know from the first problem that Marissa will spend $.90 to purchase the blue marbles in the first bag. She starts out with $1.00, so her change is the difference between $1.00 and $.90, or $.10.

The yellow marbles cost $.10 each. So, Marissa has enough change to purchase 1 yellow marble.

The correct answer is 2).

10. 5) If there are 6 games being played, that means there are 12 teams in action. If each team has the minimum of 3 fifth-graders starting, then the total number of fifth-graders starting on that day is equal to the product of 3 and 12, or 36.

The correct answer is 5).

11. 4) To help us answer the questions, it helps to mark the equal angles and the proportional sides on the similar triangles:

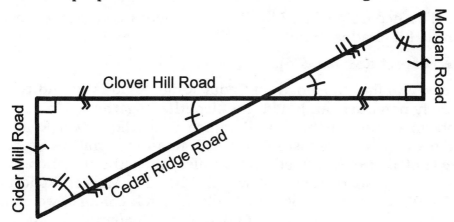

The other angle that is the size as the intersection of Cider Mill and Cedar ridge is at the upper right of the figure, at the intersection of Cedar Ridge and Morgan Road.

The correct answer is 4).

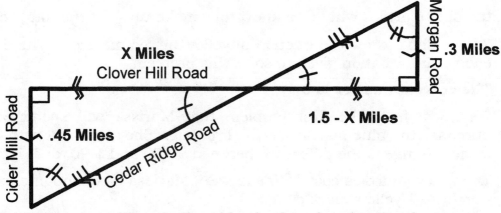

12. 4)   The ratios of the similar sides in the triangles are the same.

$$\frac{\text{Short Side Triangle 1}}{\text{Short Side Triangle 2}} = \frac{.45}{.3} = \frac{X}{(1.5 - X)} = \frac{\text{Middle Side Triangle 1}}{\text{Middle Side Triangle 2}}$$

Cross multiplying, we get:

$.675 - .45X = .3X$

Solving for X, we get

$.675 = .75X; \quad X = .9$

The distance from Cider Mill Road to where Cedar Ridge Road crosses Clover Hill Road is .9 miles.

The correct answer is 4).

13.   One way to find the length of Cedar Ridge Road is to find the length of each hypotenuse and add the lengths together. In the previous problem we found that the distance from Cider Mill Road to where Cedar Ridge Road crosses Clover Hill Road is .9 miles. Since the total length of Clover Hill Road is 1.5 miles, then the distance from where Cedar Ridge Road crosses Clover Hill Road to Morgan Road is .6 miles (1.5 - .9). Since we know the 2 sides of each triangle, we can solve for the hypotenuse by using the Pythagorean Theorem.

$$\sqrt{(.45)^2 + (.9)^2} = \sqrt{.2025 + .81} = \sqrt{1.0125} = 1.0062$$

$$\sqrt{(.3)^2 + (.6)^2} = \sqrt{.09 + .36} = \sqrt{.45} = .6708$$

$1.0062 + .6708 = 1.6770$

The easier way to solve this problem is by combining the north-south distance and solving for the whole hypotenuse in one step.

$$\sqrt{(.3 + .45)^2 + (1.5)^2} = \sqrt{.75^2 + 1.5^2} = \sqrt{.5625 + 2.25} = \sqrt{2.8125}$$

$= 1.677$

1	.	6	7	7
	⊘	⊘	⊘	
⊙	●	⊙	⊙	⊙
	⓪	⓪	⓪	⓪
●	①	①	①	①
②	②	②	②	②
③	③	③	③	③
④	④	④	④	④
⑤	⑤	⑤	⑤	⑤
⑥	⑥	●	⑥	⑥
⑦	⑦	⑦	●	●
⑧	⑧	⑧	⑧	⑧
⑨	⑨	⑨	⑨	⑨

14. The midpoint is $\left(\dfrac{x_2 + x_1}{2}, \dfrac{y_2 + y_1}{2}\right) = \left(\dfrac{5 + (-1)}{2}, \dfrac{6 + 2}{2}\right) = \left(\dfrac{4}{2}, \dfrac{8}{2}\right)$

$= (2, 4)$

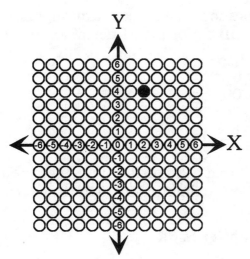

15. 3) Tina worked 5 hours Monday, 7 hours Tuesday, 6 hours Wednesday, 10 hours Thursday, 7 hours Friday, and 8 hours on Saturday. Her total number of hours for the week is:

5 + 7 + 6 + 10 + 7 + 8 = 43

Her regular pay for those hours is:

43 x $10.75 = $462.25

Her overtime pay is:

3 x .50 x $10.75 = $16.125   Rounds to $16.13

221

Her total pay is:

$462.25 + $16.13 = $478.38

The taxes withheld are:

$478.38 x .18 = $86.1084 Rounds to $86.11

Her net pay is:

$478.38 - $86.11 = $392.27

The correct answer is 3).

16. 5) If the distance between the two cities for the commute is $6\frac{2}{3}$ inches on the map, then the round trip would be represented by a distance of $13\frac{1}{3}$ inches (twice $6\frac{2}{3}$). Since each inch represents 7.5 miles, you need to find the product of $13\frac{1}{3}$ and 7.5 to determine the actual round trip distance. You can either change both numbers to improper fractions, or approximate $13\frac{1}{3}$ as a decimal.

The fraction solution is time consuming, changing each fraction to an improper fraction, canceling, if possible, and reducing your answer to lowest terms:

$$\frac{\overset{20}{\cancel{40}}}{\underset{1}{\cancel{3}}} \times \frac{\overset{5}{\cancel{15}}}{2} = 100$$

Approximating $13\frac{1}{3}$ as 13.3 and using your calculator to find the product of 13.3 and 7.5 you'll quickly get the answer 99.75. This is a close enough approximation to determine that the best response is 100 miles.

The correct answer is 5).

17. 4) If the round trip is 240 miles and George gets 20 miles/gallon in his car, by dividing 20 into 240, you'll find the number of gallons used on the round trip. 240 ÷ 20 = 12 gallons.

The correct answer is 4).

18. 3) George's mother gives him 5 cents or $.05 per mile. The total trip is 240 miles so:

240 x $.05 = $12.00

The correct answer is 3).

19. 1)   There are a couple of ways that you can look at this problem.

For one, let X represent the original price of the watch. Then, since the cost of the watch is reduced by 25% of the original cost, .25X represents the amount of reduction. The difference between the original price (X) and the reduction (.25X) is equal to the sales price ($69.00). Writing this as an equation, we get:

X - .25X = $69

.75X = $69

X = $69 ÷ .75 = $92

Another way to look at this problem is that if the price was reduced 25%, then the sales price ($69.00) must represent 75% or $\frac{3}{4}$ of the original price (X).

If $\frac{3}{4}$X = 69, then X =

$$\frac{{}^1\cancel{4}}{{}_1\cancel{3}} \cdot \frac{{}^1\cancel{3}}{{}_1\cancel{4}}X = {}^{23}\cancel{69} \cdot \frac{4}{{}_1\cancel{3}} = 92$$

By either method, we see that the original price was $92.00.

The correct answer is 1).

20.   ΔABC is a right triangle AC represents its hypotenuse. We can solve for the length of AC by using the Pythagorean Theorem.

$$\sqrt{2^2 + 2.5^2} = \sqrt{4 + 6.25} = \sqrt{10.25} \cong 3.20$$

The distance from A to C is approximately 3.2 kilometers.

		3	•	2
	⊘	⊘	⊘	
⊙	⊙	⊙	●	⊙
⓪	⓪	⓪	⓪	⓪
①	①	①	①	①
②	②	②	②	●
③	③	●	③	③
④	④	④	④	④
⑤	⑤	⑤	⑤	⑤
⑥	⑥	⑥	⑥	⑥
⑦	⑦	⑦	⑦	⑦
⑧	⑧	⑧	⑧	⑧
⑨	⑨	⑨	⑨	⑨

21. $\triangle ABC$ and $\triangle ADE$ are similar triangles and, therefore, their sides are proportional. The ratio of AB to AD is equal to the ratio of BC to DE. We also know that since B represents the midway point between points A and D, and the distance from point A to point B is 2 kilometers, then the distance from point B to point D must also be 2 kilometers. This would make the total distance between point A and point D 4 kilometers. Since AD is twice as long as AB, DE must be twice as long as BC.

Note: If you wanted to find the measurement of DE, you would set up the following proportion:

$$\frac{AB}{AD} = \frac{BC}{DE}$$

$\frac{2}{4} = \frac{2.5}{DE}$ Cross multiplying, the result is

$2(DE) = 10$ And dividing both sides by 2, the result is

$DE = 5$ The distance from point D to point E is 5 kilometers.

22. What this problem is asking for is the perimeter of the larger triangle. We know that AD equals 4 kilometers and DE equals 5 kilometers. We know that the sides of the larger triangle are twice the size of the corresponding sides of the smaller triangle. Since AC is approximately 3.2 kilometers, AE is approximately 6.4 kilometers. The perimeter of the larger triangle is equal to the sum of its three sides (4 + 5 + 6.4), or 15.4 kilometers. If 1 kilometer equals .62 miles, then 15.4 kilometers equals 15.4 times .62 miles.

15.4 × .62 = 9.548 miles. Rounded to the nearest hundredth of a mile, the result is 9.55 miles.

9	•	5	5
	Ⓘ Ⓘ Ⓘ		
⊙	⊙ ●	⊙	⊙
⓪	⓪ ⓪	⓪	⓪
①	① ①	①	①
②	② ②	②	②
③	③ ③	③	③
④	④ ④	④	④
⑤	⑤ ⑤	●	●
⑥	⑥ ⑥	⑥	⑥
⑦	⑦ ⑦	⑦	⑦
⑧	⑧ ⑧	⑧	⑧
⑨	● ⑨	⑨	⑨

23. 1) To convert F degrees Fahrenheit to C degrees Celsius, subtract 32 from F and multiply the result by $\frac{5}{9}$. As a formula this would look like:

$$C = \frac{5}{9}(F - 32).$$

The correct answer is 1).

24. 2) To convert 131 degrees Fahrenheit to Celsius, subtract 32 from 131 and multiply the result by $\frac{5}{9}$.

$$\frac{5}{9}(131 - 32)$$

$$\frac{5}{\underset{1}{9}} \cdot \overset{11}{\cancel{99}} = 55$$

131° Fahrenheit is equal to 55° Celsius.

The correct answer is 2).

25. 5) Using the formula from problem 20, but solving for F instead of C, we would get:

$C = \frac{5}{9}(F - 32)$ Multiplying both sides by $\frac{9}{5}$, the result is

$\frac{9}{5}C = F - 32$ Adding 32 to both sides, the result is

$\frac{9}{5}C + 32 = F$

The correct answer is 5).

26. 1) They paint a total of 3 rooms and each room has 4 walls, so the total number of walls painted by the pair is 12. If we let X = the number of walls that Mike painted, then the number of wall that Bob painted can be represented by 2X, since Bob painted twice as many walls as Mike. The sum of these two unknowns must equal 12, the total number of walls painted.

$$X + 2X = 12$$

$$3X = 12$$

$$X = \frac{12}{3} = 4$$

Mike painted 4 walls.

The correct answer is 1).

27. 2) If the boys paint 3 rooms in $7\frac{1}{2}$ hours, then they are working at a rate of one room every $2\frac{1}{2}$ hours. In a ten-hour workday, they will be able to complete 4 rooms.

$$10 \div 2\frac{1}{2} = \frac{10}{1} \times \frac{2}{5} = \frac{{}^{2}\cancel{10}}{1} \times \frac{2}{\cancel{5}_1} = 4$$

If they paint 4 rooms each day and work a 5 day week, then they will be able to paint 4 times 5 number of rooms.

$$4 \times 5 = 20$$

The correct answer is 2).

28. The point (-2 , 5) is found by going –2 in the X direction and 5 in the Y direction.

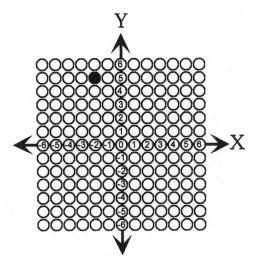

29. The area that Bob is digging out is 20' by 16', or 320 sq. ft. He is digging to a depth of 4 inches (1/3 of a foot), so the volume that he has to replace is:

$320 \times \dfrac{1}{3} = 106 \dfrac{2}{3}$ cubic feet. There are 27 cubic feet in a cubic yard, so we have to divide our answer by 27 to convert it to cubic yards.

$$106 \dfrac{2}{3} \div 27 = \dfrac{320}{3} \bullet \dfrac{1}{27} = \dfrac{320}{81} = 3 \dfrac{77}{81}$$

Since Bob has to buy whole cubic yards, he must by 4 cubic yards.

				4
	/	/	/	
•	•	•	•	•
	0	0	0	0
1	1	1	1	1
2	2	2	2	2
3	3	3	3	3
4	4	4	4	●
5	5	5	5	5
6	6	6	6	6
7	7	7	7	7
8	8	8	8	8
9	9	9	9	9

30. Cedar Mulch is $9.00 more expensive than Hardwood Bark for each cubic yard. Since Bob has to buy 4 cubic yards, for the flower beds it would be:

4 × $9.00 or $36.00 more expensive.

			3	6
	⊘	⊘	⊘	
•	•	•	•	•
	⓪	⓪	⓪	⓪
①	①	①	①	①
②	②	②	②	②
③	③	③	●	③
④	④	④	④	④
⑤	⑤	⑤	⑤	⑤
⑥	⑥	⑥	⑥	●
⑦	⑦	⑦	⑦	⑦
⑧	⑧	⑧	⑧	⑧
⑨	⑨	⑨	⑨	⑨

31. If Bob decides on Cedar Mulch for the flower beds, he will have to buy 4 cubic yards, at $25.00 per cubic yard, for a total of

4 × $25.00 or $100.00.

		1	0	0
	⊘	⊘	⊘	
•	•	•	•	•
	⓪	⓪	●	●
①	①	●	①	①
②	②	②	②	②
③	③	③	③	③
④	④	④	④	④
⑤	⑤	⑤	⑤	⑤
⑥	⑥	⑥	⑥	⑥
⑦	⑦	⑦	⑦	⑦
⑧	⑧	⑧	⑧	⑧
⑨	⑨	⑨	⑨	⑨

32.   Bob has a total of 4 cubic yards for the flower beds, or 108 cubic feet (4 × 27). If a wheelbarrow holds 3 cubic feet of mulch, divide 3 into 108 to determine the number of trips.

108 ÷ 3 = 36 trips.

		3	6
⊘	⊘	⊘	
⊙	⊙	⊙	⊙	⊙
	⓪	⓪	⓪	⓪
①	①	①	①	①
②	②	②	②	②
③	③	③	●	③
④	④	④	④	④
⑤	⑤	⑤	⑤	⑤
⑥	⑥	⑥	⑥	●
⑦	⑦	⑦	⑦	⑦
⑧	⑧	⑧	⑧	⑧
⑨	⑨	⑨	⑨	⑨

33. 4) If John does not make any Out of State calls, then the best plan for him is the one that offers the lowest rate for in-state calls. At 5¢ per minute, that would be company D.

The correct answer is 4).

34. 3) Company C and Company E offer the same rate of Off Peak State to State calls. Company E offers a better Peak rate, but John will only make Off Peak State to State calls, so this is unimportant. The only important difference is that Company C offers 250 free trial minutes. This is why John selects Company C over Company E.

The correct answer is 3).

35. 1) Plans A and E have the lowest average rate for State to State calls (7 cents per minute). However, John will select plan A because, of the two, it has the lower in-state rate.

The correct answer is 1).

36. 3) Sometimes, it's easiest to set up an example for yourself – give the rectangle your own dimensions, say 1 foot wide by 2 feet long. If you double the width, the result is 2 feet. If you triple the length, the result is 6 feet. The resulting area is 2 times 6 or 12 square feet. The area grew from 2 square feet to 12 square feet. Twelve is six times two, so the area is six times larger.

To show this relationship algebraically, we would let X represent the original width and Y represent the original length. This gives us an area of XY. Doubling the width gives us 2X and tripling the length gives us 3Y. The new area is equal to 2X times 3Y, or 6XY. Since the original area was XY, it is obvious that the new area is 6 times larger.

The correct answer is 3).

37. 3) The mode is the most frequently occurring score. The team scored the same number of points in each of the three games from October 29th through November 12th. The point on the graph in each case is just below the grid line marked "25". Even if you were uncertain as to whether this value was equal to 23 or 24, based on the 5 possible answers, it's obvious that the answer is 24.

The correct answer is 3).

38. 4) The team played four games in October. They scored 14 points on October 1st, 13 points on October 8th, 19 points on October 15th, and 24 points on October 29th. The most points they scored in any one game was 24.

The correct answer is 4).

39. 5) The team played four games in December. They scored 9 points on December 3rd, 30 points on December 10th, 17 points on December 17th, and 28 points on December 23rd. The total number of points that they scored in December is equal to the sum of the points scored in each of these four games.

$9 + 30 + 17 + 28 = 84$.

The correct answer is 5).

40. 3) Since Michelle is graded on 5 subjects each marking period and there are 4 marking periods each year, she will receive a total of 20 grades for the year. If X equal the number of A's she receives and we know that she received at least 5 A's over the course of the year, then the two formulas could be represented as:

$4.00 for every "A" she earns would be represented by 4X.

(X – 5) represents the number of A's in excess of 5 that she earns. If she earns $3.00 for the first 5 A's and $5.00 for each "A" in excess of 5, this second formula would be represented as $3(5) + 5(X – 5)$.

So, the two formulas are 4X and $3(5) + 5(X – 5)$.

The correct answer is 3).

41. 3)   If Michelle is to receive the same amount of money under either formula, we would set the two equal to each other and solve for X.

$$4X = 3(5) + 5(X - 5)$$

$$4X = 15 + 5X - 25$$

$$-X = -10$$

$$X = 10$$

Ten A's will earn Michelle 4(10) or $40.00 under the first formula, and she'll earn 3(5) + 5(10 − 5), or $40.00 under the second formula.

The correct answer is 3).

42. 5)   If X represents the total length in minutes, then (X − 20) represents the number of minutes in excess of 20 minutes. The first 20 minutes are charged at a flat rate of $1.00. The remaining minutes (X − 20), are charged at 7 cents each. The total charge for the minutes in excess of 20 is the product of the rate ($.07) and the number of minutes (X − 20), which can be represented as $.07(X − 20). The total cost of the call can be represented as the sum of the flat rate ($1.00) and the cost of the minutes in excess of 20, ($.07(X − 20))

or 1 + .07(X − 20).

The correct answer is 5).

43. 4)   The break even point between the two plans, is when the cost of the call under the first plan

$$1 + .07(X - 20)$$

equals the flat rate under the second plan

.06X

Setting up an equation and solving for X, we find that

$$1 + .07(X - 20) = .06X$$

$$1 + .07X - \$1.4 = .06X$$

$$-.4 = -.01X$$

$$40 = X$$

The break even point is 40 minutes.

The correct answer is 4).

232

44. 5) A person would pay $1.00 for the first 20 minutes and .07(27) for the next 27 minutes.

1.00 + .07(27) = 1.00 + 1.89 = 2.89

The cost would be $2.89.

The correct answer is 5).

45. 4) The highest bar occurs in the month of August.

The correct answer is 4).

46. 3) The bar for February ends just above the 400 mark. January and November are below the 400 mark, and March appears to be directly on the 400 mark. All the other bars are above the 500 mark except for December. December is between 400 and 500, but definitely larger than February. So, there are 3 months (January, March, and November) where there was less electric usage than February.

The correct answer is 3).

47. 4) For each bar that doesn't fall directly on a marked line, you have to approximate the value. The possible answers are spread out sufficiently so that if you make good estimates, there should be no problem coming up with the correct response. Remember, the values below are just approximations.

Jan	325
Feb	420
Mar	400
Apr	575
May	625
Jun	750
Jul	875
Aug	1000
Sep	850
Oct	550
Nov	380
Dec	450
	7200

The correct answer is 4).

48. 1) Since the diameter of a circle is twice the radius, the formula for circumference could also be represented as $C = 2\pi r$. The ratio of Area to Circumference is:

$$\frac{\pi r^2}{2\pi r}$$

The factors of π in numerator and denominator will cancel each other out. So will one factor of r, leaving one factor of r in the numerator. Reducing the ratio to lowest terms, we get:

$$\frac{r}{2}$$

The correct answer is 1).

49. 3) $A = \pi r^2$. To see what happens to the area when we triple the radius, we substitute 3r for r in the formula. The key to success here is remembering that the entire expression "3r" is squared.

$A = \pi(3r)^2$

$A = \pi(9r^2)$

$A = 9\pi r^2$

When you triple the radius the area becomes 9 times larger.

The correct response 3)

50. 1) $C = \pi d$, or $C = 2\pi r$. If we double the radius and what to see what happens to the Circumference, we substitute 2r for r in the formula.

$C = 2\pi(2r)$

$C = 4\pi r$

The circumference grows from $2\pi r$ to $4\pi r$. When we double the radius, the circumference also doubles.

The correct answer is 1).

Appendix A: Using the Calculator

The Mathematics section of the GED exam is divided into two parts, each with an equal number of questions (25) and time limit (45 minutes). For part one, you will be allowed to use a calculator. The calculator will be issued to you at the testing center. It is important that you are familiar with the official calculator, CASIO® fx-260 SOLAR.

It is also important that you are able to locate and use the following keys:

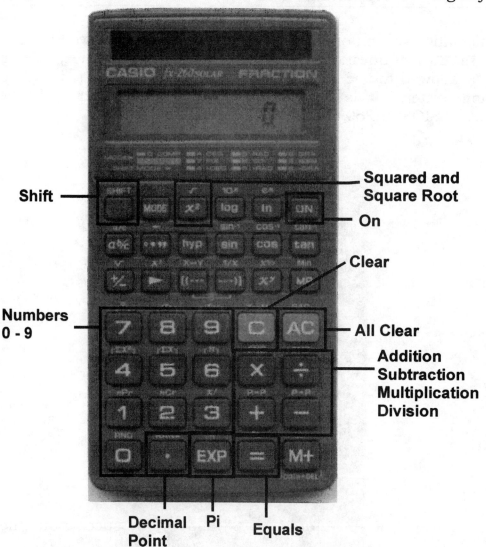

Shift

Squared and Square Root

On

Clear

Numbers 0 - 9

All Clear

Addition Subtraction Multiplication Division

Decimal Point

Pi

Equals

Example: To perform the following calculation, 27 + 64, on the calculator, you would first press the numbers "2" and "7" in succession (the number 27 will appear in the display area at the top of the calculator), then press the "+" key, then press the numbers "6" and "4" in succession (the number 64 will appear in the display area at the top of the calculator), and finally, press the "=" key. The number 91, the result of adding 27 and 64 together, will appear in the display area at the top of the calculator.

Example: To perform the following calculation, 105 - 39, on the calculator, you would first press the numbers "1", "0", and "5" in succession (the number 105 will appear in the display area at the top of the calculator), then press the "-" key, then press the numbers "3" and "9" in succession (the number 39 will appear in the display area at the top of the calculator), and finally, press the "=" key. The number 66, the result of subtracting 39 from 105, will appear in the display area at the top of the calculator.

Example: To perform the following calculation, 79 × 3.5, on the calculator, you would first press the numbers "7" and "9" in succession (the number 79 will appear in the display area at the top of the calculator), then press the "×" key, then press the number "3", the decimal point ("."), and the number "5" in succession (the number 3.5 will appear in the display area at the top of the calculator), and finally, press the "=" key.   The number 276.5, the result of multiplying 79 from 3.5, will appear in the display area at the top of the calculator.

Example: To perform the following calculation, 408 ÷ 48, on the calculator, you would first press the numbers "4", "0" and "8" in succession (the number 408 will appear in the display area at the top of the calculator), then press the "÷" key, then press the numbers "4" and "8" in succession (the number 48 will appear in the display area at the top of the calculator), and finally, press the "=" key.   The number 8.5, the result of dividing 408 from 48, will appear in the display area at the top of the calculator.

Example: To perform the following calculation, $(51.25)^2$, on the calculator, you would first press the numbers "5", "1", the decimal point ("."), and the numbers "2" and "5" in succession (the number 51.25 will appear in the display area at the top of the calculator), then press the "x^2" key.   The number 2,626.5625, the square of 51.25, will appear in the display area at the top of the calculator.

Example: To perform the following calculation, $\sqrt{163.84}$, on the calculator, you would first press the numbers "1", "6", "3", the decimal point ("."), and the numbers "8" and "4" in succession (the number 163.84 will appear in the display area at the top of the calculator).

Key Point

This is the tricky part of using this calculator. In order to perform the square root function you must hit two keys.

press the "Shift" key, and then press the "x^2" key.   The number 12.8, the square root of 163.84, will appear in the display area at the top of the calculator.

By pressing the "Shift" key first, the calculator will execute the command that is written **above** the subsequent key.   The command written above the "x^2" key is " $\sqrt{}$ ".

There are a few other "administrative" keys that will be helpful for you to remember.

Key Point

If you are entering a number, and you haven't hit a function key yet, you can clear the display by hitting the "Clear" key.

Key Point

By hitting the "All Clear" key you will clear the display of all work, including any functions that have been performed. This prepares the calculator for a new calculation that will not be linked to any previous work.

The last key you can try using after you have mastered all the other keys is the backspace key. It can save you time when compared to the clear key if you realize your mistake immediately.

Key Point

If you are entering a multi-digit number and make a mistake on entering one of the digits, you do not have to clear and start over. You can use the "Backspace" key to erase only the incorrect digit.

"Alternate format questions" is just a fancy way of saying that the question is not a multiple-choice question. There are going to be at least two types on the mathematics section of the GED test. The two most common types are the grid and the coordinate plane grid. This section will give you a brief introduction to both.

Grid Questions

Below is an example of a grid.

Let's take a look at each part of the grid so that you will be able to answer this type of question easily. Let's start at the top.

The first row contains empty boxes. These are for your use, but will not be looked at when your test is corrected. You can write your answer in the boxes. This will help you grid in your answer. You will be able to look back up at your columns to compare the number or function you filled in with the number you wrote at the top of the column. If they do not match you filled in the wrong oval. Erase the incorrect one and fill in the correct one.

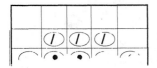

The second row contains three slash marks. These are for entering fractions. The two side columns do not contain slashes because there would be no way to enter the numerator or denominator with those columns filled in. The numbers to the left of the slash would be the numerator, and the numbers to the right would be the denominator.

239

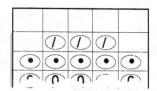

The third row of the grid contains a series of five decimal points. Use these to enter your decimal numbers.

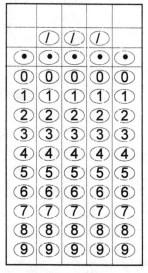

The remaining ten rows contain the numbers from 0 to 9 and they are used to enter the numerical numbers in your answer.

It is important to note that all five columns do not have to be filled in. For instance if your answer contains only three digits (or two digits and a fraction slash or decimal point), you would have two blank columns. They could both be on the left, both be on the right, or you could center your answer and put one on either side. They will all be scored the same way. So that you do not confuse yourself you should consistently leave the blank columns either on the left or the right. Let's look at some examples.

Example: Grid in $\frac{3}{4}$.

All three of the grids would be correct. From now on we will only show the grids right justified (blank columns to the left).

Example: Grid in .757

Example: Grid in 99.3

Coordinate Grid Problems

The second type of alternate format question is the coordinate grid question. In this type of question you will be required to enter in one specific point. If more than one circle is darkened in the question will be marked wrong.

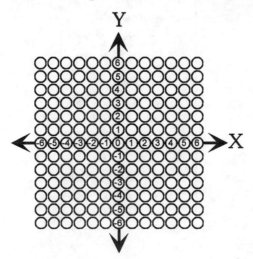

The grid consists of a box of thirteen columns and rows of circles. The x-axis is marked on the middle row and the circles in that row are numbered from –6 to +6. The middle column is labeled the y-axis and the circles are labeled from +6 to –6. This means that you will be able to darken in the points (-6,-6) to (6,6). Let's look at some examples.

Example: Grid in the point (3,5)

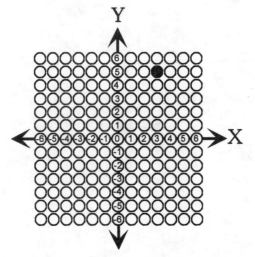

You would move from the center circle labeled with a "0" three spaces to the right in an x direction and then five spaces up in the y direction.

Example: Grid in the point (-2,-3)

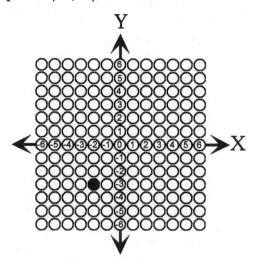

You would move from the center circle labeled with a "0" two spaces to the left (the negative direction) in an x direction and then three spaces down (the negative direction) in the y direction.

With a little practice the alternate format questions will be as easy to answer as any other questions.

NOTES

NOTES

NOTES